Geometric Dimensioning and Tolerancing

by

David A. Madsen

Department Chairperson—Drafting Technology
Clackamas Community College
Oregon City, Oregon

Autodesk Authorized Training Center® for AutoCAD®
Softdesk Authorized Training Center for Civil Engineering

Past Member Board of Directors
American Design Drafting Association (ADDA)
ADDA Drafter and Design Drafter Certified Curricula

South Holland, Illinois
The Goodheart-Willcox Company, Inc.
Publishers

Library of Congress Catalog Card Number 94-17779
International Standard Book Number 1-56637-064-7

3 4 5 6 7 8 9 10   95   99 98 97 96 95

Library of Congress Cataloging-in-Publication Data

Madsen, David A.
   Geometric dimensioning and tolerancing / by David A. Madsen.
     p.   cm.
   Includes index.
   ISBN 1-56637-064-7
   1. Engineering drawings–Dimensioning.   2. Tolerance (Engineering)
I. Title.
T357.M22   1994
604.2′43–dc20   94-17779
                                                 CIP

Content Notice

This book contains the most complete and accurate information that could be obtained from various authoritative sources at the time of publication. Goodheart-Willcox cannot assume responsibility for any changes, errors, or omissions.

Introduction

Geometric Dimensioning and Tolerancing provides you with the complete fundamentals of geometric dimensioning and tolerancing (GD&T) concepts as adopted by the American National Standards Institute (ANSI) and published by The American Society of Mechanical Engineers (ASME) for engineering and related documentation practices. The standard is titled ANSI/ASME Y14.5M-1994 *Dimensioning and Tolerancing*. ANSI/ASME Y14.5M-1994 is a revision of ANSI Y14.5M-1982. In this text, this standard will be referred to as ASME Y14.5M. Other Y14.5 documents include ASME Y14.5.1 *Mathematical Definition of Y14.5* and ASME Y14.5.2 *Certification of GD&T Professionals*. The standards documents that control general dimensional tolerances found in the title block and in general notes is ANSI Y14.1 *Drawing Sheet Size and Format* for inch applications and ASME Y14.1M *Metric Drawing Sheet Size and Format* for metric applications. Engineering change applications are recommended by the document ASME Y14.35M *Revision of Engineering Drawings and Associated Documents*. The American Society of Mechanical Engineers is located at 345 East 47th Street, New York, NY 10017.

PURPOSE

The goal of this workbook is to guide you through a logical sequence of learning activities and to use this knowledge in a skill building format.

❏ It is important for you to have a solid foundation in the understanding of dimensioning and tolerancing terms, definitions, and concepts before beginning a study of geometric dimensioning and tolerancing.

❏ The geometric dimensioning and tolerancing concepts are introduced to you in a methodical manner to help ensure that you have full understanding of every basic concept as you build knowledge toward more advanced applications.

❏ The concepts are covered in an easy-to-learn sequence.

❏ Learning about GD&T progresses using a format that allows you to become comfortable with the concepts as you build understanding from one chapter to the next.

❏ The basics of reading and drawing geometric dimensioning and tolerancing should be mastered before advanced topics such as designing and inspecting for GD&T are covered.

❏ GD&T is approached as an easy to understand subject rather than something complex and alien.

To the Student

Geometric Dimensioning and Tolerancing is for Manufacturing Technology students, Drafting Technology students, and for professional upgrade training. Everything you need is in this workbook.

For Manufacturing Technology students:
- ❏ Chapter tests reinforce the previously learned topics.
- ❏ Print reading exercises are provided following every chapter. This gives you the opportunity to read prints containing the geometric dimensioning and tolerancing information related to the chapter.
- ❏ The prints are actual industry drawings that were created using computer-aided design and drafting (CADD).
- ❏ Computer-aided manufacturing (CAM) concepts are introduced.
- ❏ A comprehensive final exam is provided to solidify the learning that has been conducted throughout the course of study.

For Drafting Technology students:
- ❏ Chapter tests reinforce the previously learned topics.
- ❏ The print reading exercises are optional, but may be considered helpful in understanding more about the information provided in prints.
- ❏ Numerous drafting problems are presented as 3-D illustrations or design layouts. This requires that you determine the correct views, dimension, and geometric dimensioning and tolerancing placement.
- ❏ Drafting problems range from basic to advanced.
- ❏ Drafting problems may be completed using manual or computer-aided drafting. GD&T is a natural for CADD applications. This book explains how GD&T relates to using a CADD system.
- ❏ Symbols are detailed throughout the book to demonstrate proper drafting standards.
- ❏ Drafting rules and standards are emphasized for proper applications.
- ❏ A comprehensive final exam is provided to solidify the learning that has been conducted throughout the course of study.

ACKNOWLEDGEMENTS

I would like to give special thanks to the people who gave professional technical support for this edition of *Geometric Dimensioning and Tolerancing*:

AutoCAD® illustrations for the text examples:
Keith McDonald
Mark and Wayne Niemeyer
Earl Larson

AutoCAD® illustrations for the drawing problems:
Joe Ballweber
Eric Lewis

Industry support:
Richard Phillips, Documentation Supervisor, FLIR Systems, Inc.
Keith McDonald, FLIR Systems, Inc.
George Schafer, Manager Engineering Services, NACCO Material Handling Inc.
Gary Whitmire, ASME Standards Committee Y14.
James R. Larson, Ingersoll Milling Machine Company

Cover drawing:
Courtesy of FLIR Systems, Inc.

—**David A. Madsen**

Contents

Chapter 3
Datums _____ 61

Chapter 4
Material Condition Symbols _____ 99

Chapter 5
Tolerances of Form and Profile _____ 127

Prints for End-of-Chapter Print Reading Exercises _____ 303

Appendices _____ 317

Glossary _____ 335

Index _____ 345

Chapter 1

Dimensioning and Tolerancing

This chapter covers general tolerancing as applied to conventional dimensioning practices. The term *conventional dimensioning* as used here implies dimensioning without the use of geometric tolerancing. *Conventional tolerancing* applies a degree of form and location control by increasing or decreasing the tolerance.

Conventional dimensioning methods provide the necessary basic background to begin a study of geometric tolerancing. It is important that you completely understand conventional tolerancing before you begin the study of geometric tolerancing.

When mass production methods began, interchangeability of parts was important. However, many times parts had to be "hand selected for fitting." Today, industry has faced the reality that in a technological environment, there is no time to do unnecessary individual fitting of parts. Geometric tolerancing helps ensure interchangeability of parts. The function and relationship of a particular feature on a part dictates the use of geometric tolerancing.

Geometric tolerancing does not take the place of conventional tolerancing. However, geometric tolerancing specifies requirements more precisely than conventional tolerancing, leaving no doubts as to the intended definition. This precision may not be the case when conventional tolerancing is used and notes on the drawing may become ambiguous.

When dealing with technology, a drafter needs to know how to properly represent conventional dimensioning and geometric tolerancing. Also, a technician must be able to accurately read dimensioning and geometric tolerancing. Generally, the drafter converts engineering sketches or instructions into formal drawings using proper standards and techniques. After acquiring adequate experience, a design drafter, designer, or engineer begins implementing geometric dimensioning and tolerancing on the research and development of new products or the revision of existing products.

Most dimensions in this text are in metric. Therefore, a 0 precedes decimal dimensions less than one millimeter, as in 0.25. When inch dimensions are used, a 0 will *not* precede a decimal dimension that is less than one inch.

DIMENSIONING UNITS

Most dimensions in this text are the metric International System of Units (SI). Separate problems and print reading exercises are provided with metric and inch dimensions. The common SI unit of measure used on engineering drawings is the millimeter. The common U.S. unit used on engineering drawings is the inch. The actual units used on your engineering drawings will be determined by the policy of your school or company. The general note "UNLESS OTHERWISE SPECIFIED, ALL DIMENSIONS ARE IN MILLIMETERS" (or "INCHES") should be placed on the drawing when all dimensions are in either millimeters or inches. When some inch dimensions are placed on a metric drawing, the abbreviation "IN." should follow the inch dimensions. The abbreviation "mm" should follow any millimeter dimensions on a predominately inch-dimensioned drawing. Angular dimensions are established in degrees (°) and decimal degrees (X.X°), or in degrees (°) minutes (') and seconds (").

The following are some rules for metric and inch dimension units. Examples of these rules are shown in Example 1-1.

Millimeters

❑ The decimal point and zero are omitted when the metric dimension is a whole number. For example, the metric dimension "12" has no decimal point followed by a zero.

❑ When the metric dimension is greater than a whole number by a fraction of a millimeter, the last digit to the right of the decimal point is not followed by a zero. For example, the metric dimension "12.5" has no zero to the right of the five. This rule is true unless tolerance values are displayed.

❑ Both the plus and minus values of a metric tolerance have the same number of decimal places. Zeros are added to fill in where needed.

❑ A zero precedes a decimal millimeter that is less than one. For example, the metric dimension "0.5" has a zero before the decimal point.

❑ Examples in ASME Y14.5M show no zeros after the specified dimension to match the tolerance. For example, 24±0.25 or 24.5±0.25 are correct. However, some companies prefer to add zeros after the specified dimension to match the tolerance, as in 24.00±0.25 or 24.50±0.25.

Inch

❑ A zero does *not* precede a decimal inch that is less than one. For example, the inch dimension ".5" has no zero before the decimal point.

❑ A specified inch dimension is expressed to the same number of decimal places as its tolerance. Zeros are added to the right of the decimal point if needed. For example, the inch dimension ".250±.005" has an additional zero added to ".25" to match the three decimal tolerance.

❑ Fractional inches may be used, but generally indicate a larger tolerance. Fractions may be used to give nominal sizes such as in a thread callout.

❑ Both the plus and minus values of an inch tolerance have the same number of decimal places. Zeros are added to fill in where needed.

❑ Zeros are added where needed after the specified dimension to match the tolerance. For example, 2.000±.005 and 2.500±.005 both have zeros added to match the tolerance.

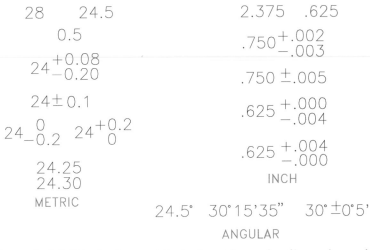

Example 1-1. Displaying metric, inch, and angular dimension units.

FUNDAMENTAL DIMENSIONING RULES

The following rules are summarized from ASME Y14.5M. These rules are intended to give you an understanding of the purpose for standardized dimensioning practices. Short definitions are given in some cases for terminology that is explained in detail later in this text.

❑ Each dimension has a tolerance except for dimensions specifically identified as reference, maximum, minimum, or stock. The tolerance may be applied directly to the dimension, indicated by a general note, or located in the title block of the drawing.

❑ Dimensioning and tolerancing must be complete to the extent that there is full understanding of the characteristics of each feature. Neither measuring the drawing or assumption of a dimension is permitted. Exceptions include drawings such as loft, printed wiring, templates, and master layouts prepared on stable material. However, in these cases the necessary control dimensions must be given.

❑ Each necessary dimension of an end product must be shown. Only dimensions needed for complete definition should be given. Reference dimensions should be kept to a minimum.

❑ Dimensions must be selected and arranged to suit the function and mating relationship of a part. Dimensions must not be subject to more than one interpretation.

❑ The drawing should define the part without specifying the manufacturing processes. For example, give only the diameter of a hole without a manufacturing process such as "DRILL" or "REAM." However, there should be specifications given on the drawing, or related documents, in cases where manufacturing, processing, quality assurance, or environmental information is essential to the definition of engineering requirements.

❑ It is allowed to identify (as nonmandatory) certain processing dimensions that provide for finish allowance, shrink allowance, and other requirements, provided the final dimensions are given on the drawing. Nonmandatory processing dimensions should be identified by an appropriate note, such as "NONMANDATORY (MFG DATA)."

❑ Dimensions should be arranged to provide required information arranged for optimum readability. Dimensions should be shown in true profile views and should refer to visible outlines.

❏ Wires, cables, sheets, rods, and other materials manufactured to gage or code numbers should be specified by dimensions indicating the diameter or thickness. Gage or code numbers may be shown in parentheses following the dimension.

❏ A 90° angle is implied where centerlines, and lines displaying features, are shown on a drawing at right angles and no angle is specified. The tolerance for these 90° angles is the same as the general angular tolerance specified in the title block or in a general note.

❏ A 90° basic angle applies where centerlines of features are located by basic dimensions and no angle is specified. *Basic dimensions* are considered theoretically perfect in size, profile, orientation, or location. Basic dimensions are the basis for variations that are established by other tolerances.

❏ Unless otherwise specified, all dimensions are measured at 20°C (68°F). Compensation may be made for measurements taken at other temperatures.

❏ All dimensions and tolerances apply in a free state condition except for nonrigid parts. *Free state condition* describes distortion of the part after removal of forces applied during manufacturing. *Nonrigid parts* are those that may have dimensional change due to thin wall characteristics.

❏ Unless otherwise specified, all geometric tolerances apply for full depth, length, and width of the feature.

❏ Dimensions apply on the drawing where specified.

DEFINITIONS RELATED TO TOLERANCING

A review of the following definitions is suggested to help you gain a good understanding of the terminology associated with dimensioning practices. Additional terminology is provided as you continue through this chapter.

Actual size: The measured size of a feature or part after manufacturing.

Diameter: The distance across a circle measured through the center. Represented on a drawing with the symbol "∅" as shown in Example 1-2. Circles on a drawing are dimensioned with a diameter.

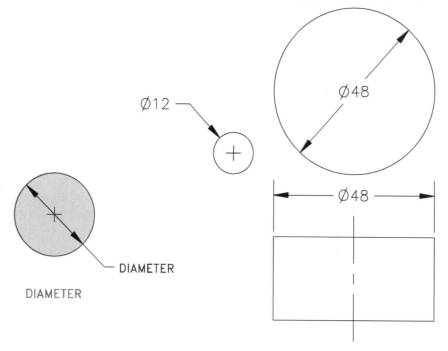

Example 1-2. Dimensioning circles with a diameter.

Dimension: A numerical value indicated on a drawing and in documents to define the size, shape, location, geometric characteristics, or surface texture of a feature. Dimensions are expressed in appropriate units of measure.

Feature: The general term applied to a physical portion of a part or object. A surface, slot, tab, keyseat, or hole are all examples of features.

Feature of size: One cylindrical or spherical surface, or a set of two parallel plane surfaces, each feature being associated with a size dimension.

Nominal size: A dimension used for general identification such as stock size or thread diameter.

Radius: The distance from the center of a circle to the outside. Arcs are dimensioned on a drawing with a radius. A radius dimension is preceded by an "R." The symbol "CR" refers to a controlled radius. *Controlled radius* means that the limits of the radius tolerance zone must be tangent to the adjacent surfaces, and there can be no reversal in the contour. The use of CR is more restrictive than R (where reversals are permitted). See Example 1-3. The symbol "SR" refers to a *spherical radius.*

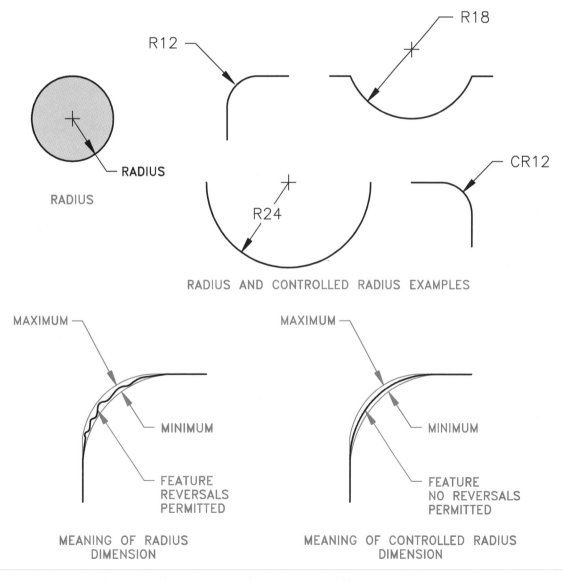

Example 1-3. Dimensioning a radius and a controlled radius.

Reference dimension: A dimension, usually without a tolerance, used for information purposes only. This dimension is often a repeat of a given dimension or established from other values shown on the drawing. A reference dimension does *not* govern production or inspection. A reference dimension is shown on a drawing with parentheses. For example, (60) would indicate a reference dimension.

Stock size: A commercial or premanufactured size, such as a particular size of square, round, or hex steel bar.

TOLERANCING FUNDAMENTALS

A *tolerance* is the total amount that a specific dimension is permitted to vary. A tolerance is not given to values that are identified as reference, maximum, minimum, or stock sizes. The tolerance may be applied directly to the dimension, indicated by a general note, or identified in the drawing title block. Refer to Example 1-4.

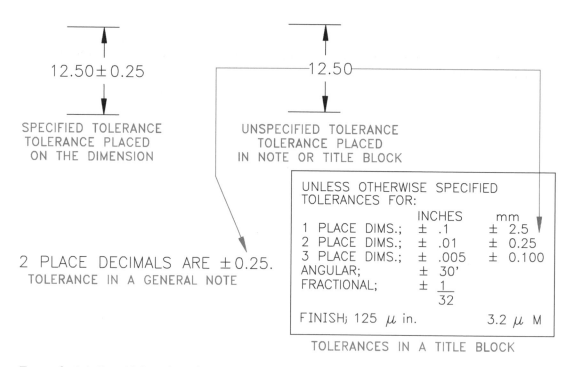

Example 1-4. Specifying the tolerance on the dimension, in a general note, or in the drawing title block.

The *limits* of a dimension are the largest and smallest numerical value that the feature can be. In Example 1-5a, the dimension is stated as 12.50±0.25. This is referred to as *plus-minus dimensioning*. The tolerance of this dimension is the difference between the maximum and minimum limits. The upper limit is 12.50 + 0.25 = 12.75 and the lower limit is 12.50 – 0.25 = 12.25. So, if you take the upper limit and subtract the lower limit you have the tolerance: 12.75 – 12.25 = 0.50.

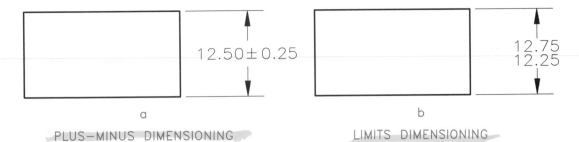

Example 1-5. Plus-minus dimensioning and limit dimensioning.

The *specified dimension* is the part of the dimension from where the limits are calculated. The specified dimension of the feature shown in Example 1-5 is 12.5. A dimension on a drawing may be displayed with plus-minus dimensioning, or the limits may be calculated and shown as in Example 1-5b. Many companies prefer this second method because the limits are shown and calculations are not required. This is called *limits dimensioning*.

A *bilateral tolerance* is permitted to vary in both the + and the – directions from the specified dimension. An *equal bilateral tolerance* is where the variation from the specified dimension is the same in both directions. An *unequal bilateral tolerance* is where the variation from the specified dimension is not the same in both directions. Refer to Example 1-6.

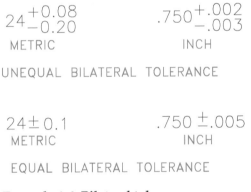

Example 1-6. Bilateral tolerances.

A *unilateral tolerance* is permitted to increase or decrease in only one direction from the specified dimension. Refer to Example 1-7.

Example 1-7. Unilateral tolerances.

MAXIMUM MATERIAL CONDITION (MMC)

Maximum Material Condition is the condition where a feature of size contains the maximum amount of material within the stated limits. The key words are "maximum amount of material."

An external feature is at Maximum Material Condition at its largest limit, or maximum amount of material, as shown in Example 1-8.

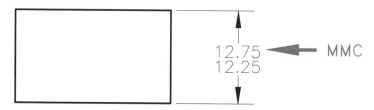

Example 1-8. Maximum Material Condition (MMC) for external features.

An internal feature is at Maximum Material Condition at its smallest limit, or maximum amount of material, as shown in Example 1-9.

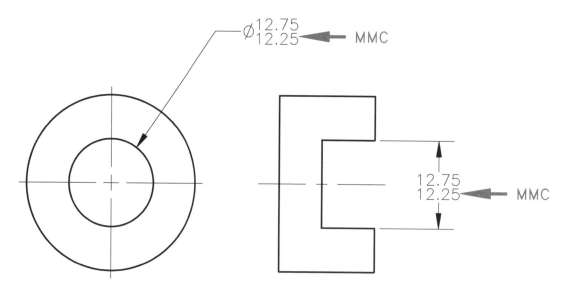

Example 1-9. Maximum Material Condition (MMC) for internal features.

LEAST MATERIAL CONDITION (LMC)

Least Material Condition is the condition where a feature of size contains the least amount of material within the stated limits. The key words are "least material."

An external feature is at Least Material Condition at its smallest limit, or least amount of material, as illustrated in Example 1-10.

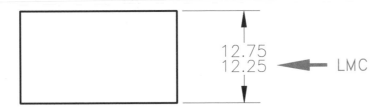

Example 1-10. Least Material Condition (LMC) for external features.

An internal feature is at Least Material Condition at its largest limit, or least amount of material, as shown in Example 1-11.

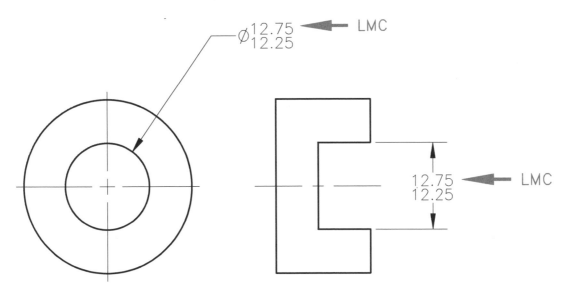

Example 1-11. Least Material Condition (LMC) for internal features.

EXTREME FORM VARIATION

The limits of size of a feature controls the amount of variation in size and geometric form. This is referred to as "Rule 1" in ASME Y14.5M. The *limits of size* is the boundary between MMC and LMC. The form of the feature may vary between the upper limit and lower limit of a size dimension. This is known as *extreme form variation*, as shown in Example 1-12.

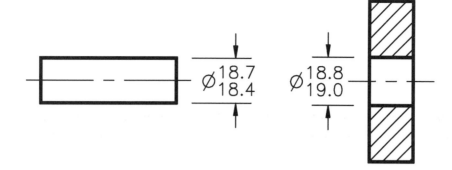

THE DRAWING

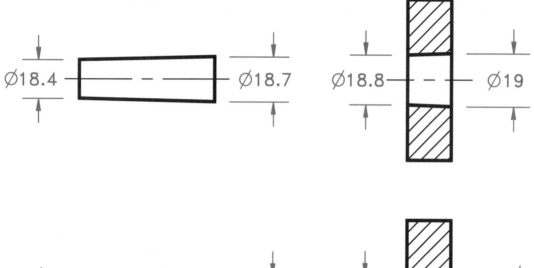

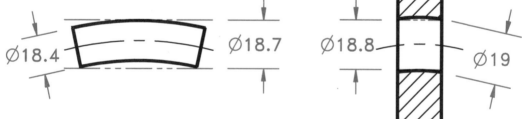

THE MEANING
EXTREME FORM VARIATIONS

Example 1-12. Extreme form variation.

The control of geometric form established by limits of size does not apply to premanufactured items such as stock bars, sheets, tubing, or structural shapes. These items are produced to government or industry standards that have established geometric tolerances such as straightness and flatness. These standards govern cases where the finished product contains the original premanufactured shape, unless other geometric tolerances are specified on the drawing.

BASIC FITS OF MATING PARTS

Standard ANSI Fits

Based on ANSI B4.1, the three general groups of limits and fits between mating parts are: running and sliding fits, force or interference fits, and locational fits.

Running and sliding fits (RC) are intended to provide a running performance with suitable lubrication allowance. Running fits range from RC 1 (close fits) to RC 9 (loose fits).

Force fits (FN) or *shrink fits* constitute a special type of interference fit characterized by maintenance of constant pressure. Force fits range from FN 1 (light drive) to FN 5 (force fits required in high stress applications).

Locational fits are intended to determine only the location of the mating parts.

Standard ANSI/ISO Fits

Based on ANSI B4.2, the general groups of metric limits and fits between mating parts are: clearance fits, transition fits, and interference fits. Example 1-13 shows the ISO symbols used to represent metric fits and gives a description of the different metric fits.

Clearance fits are generally the same as the running and sliding fits explained previously. With clearance fits, a clearance exists between the mating parts under all tolerance conditions.

Transition fits may result in either a clearance fit or an interference fit due to the range of limits between mating parts.

Interference fits require that the mating parts be pressed or forced together under all tolerance conditions.

TYPE OF FIT	ISO SYMBOL		DESCRIPTION OF FIT
	HOLE	SHAFT	
CLEARANCE FIT	H11/c11	C11/H11	LOOSE RUNNING
	H9/d9	D9/h9	FREE RUNNING
	H8/f8	F8/h7	CLOSE RUNNING
	H7/g6	G7/h6	SLIDING
	H7/h6	H7/h6	LOCATIONAL CLEARANCE
TRANSITION FIT	H7/k6	K7/h6	LOCATIONAL TRANSITION
	H7/n6	N7/h6	LOCATIONAL TRANSITION
INTERFERENCE FIT	H7/p6'	P7/h6	LOCATIONAL INTERFERENCE
	H7/s6	S7/h6	MEDIUM DRIVE
	H7/u6	U7/h6	FORCE

Example 1-13. ISO symbols used to represent metric fits and a description of these fits.

For more information on fits refer to the *Machinery's Handbook* under the classification of ALLOWANCES AND TOLERANCES FOR FITS, ANSI STANDARD FITS, GRAPHICAL REPRESENTATION OF LIMITS AND FITS, and STANDARD FIT TABLES.

CLEARANCE FIT

A clearance fit is shown in Example 1-14. In that illustration, Part 1 fits into Part 2 with a clearance between the two parts no matter what the actual size of each part is when produced within the given tolerances.

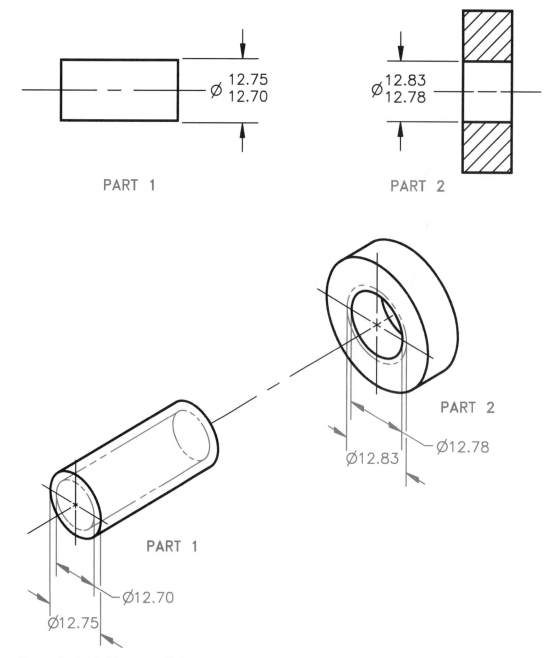

Example 1-14. Clearance fit between two parts.

ALLOWANCE

Allowance is defined as an intentional difference between the maximum material limits of mating parts. Allowance is the minimum clearance (positive allowance), or maximum interference (negative allowance) between mating parts. Allowance can be considered to be the tightest possible fit between parts. Allowance can be calculated using the formula:

MMC HOLE
– MMC SHAFT
ALLOWANCE

Now refer back to Example 1-14 as you make these calculations:

MMC HOLE (Part 2) = 12.78
– MMC SHAFT (Part 1) = 12.75
ALLOWANCE = 0.03

CLEARANCE

The loosest fit or maximum intended difference between mating parts is called the *clearance*. The clearance is calculated with this formula:

LMC HOLE
– LMC SHAFT
CLEARANCE

Refer again to Example 1-14 as you determine the clearance:

LMC HOLE (Part 2) = 12.83
– LMC SHAFT (Part 1) = 12.70
CLEARANCE = 0.13

FORCE FIT

A *force fit* is also referred to as an *interference fit* or a *shrink fit.* This is where two mating parts must be pressed or forced together. Due to the tolerance on each part, the shaft is larger than the hole, as shown in Example 1-15. At any produced size within the stated tolerance, the shaft will be larger than the hole.

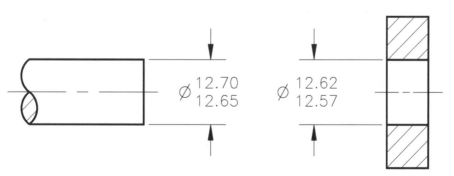

Example 1-15. Force fit between two parts.

The smallest amount of interference is:

LMC SHAFT = 12.65
– LMC HOLE = 12.62
MIN INTERFERENCE = 0.03

The greatest amount of interference is:

MMC SHAFT = 12.70
– MMC HOLE = 12.57
MAX INTERFERENCE = 0.13

CHAIN VS DATUM DIMENSIONING

The difference between chain and datum dimensioning is shown in Example 1-16. In chain dimensioning, each dimension is dependent on the previous dimension. In datum dimensioning, each dimension stands alone. Caution should be used when chain dimensioning because the tolerance of each dimension builds on the next. This is referred to as **tolerance buildup** or **stacking**. An example of tolerance buildup is when three chain dimensions have individual tolerances of ±0.2 and each feature is manufactured at or near the +0.2 limit. The potential tolerance buildup is 3 × 0.2 for a total of 0.6. To accommodate this buildup the overall dimension must have a tolerance of +0.6. This problem does not occur with datum dimensioning because each dimension is independent.

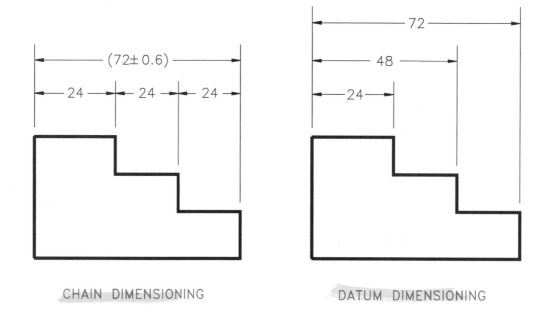

CHAIN DIMENSIONING   DATUM DIMENSIONING

UNSPECIFIED TOLERANCES ARE ± 0.2

Example 1-16. The difference between chain and datum dimensioning.

ALTERNATE DIMENSIONING PRACTICES

In industries where computer numerical control (CNC) machining or computer-aided manufacturing (CAM) processes are used, it is becoming a practice to omit dimensioning lines. This type of dimensioning is called *arrowless dimensioning* and *arrowless tabular dimensioning.*

Where changing values of a feature are involved, dimensions may be displayed in a chart. This is referred to as *chart dimensioning.*

Arrowless Dimensioning

Arrowless dimensioning is often used on precision sheet metal fabrication drawings. This type of dimensioning provides only extension lines and numbers. All dimension lines and arrowheads are omitted. Dimension numbers are aligned with the extension lines. The dimensions are read from the bottom of the drawing sheet for horizontal dimensions and from the right side for vertical dimensions. Each dimension number represents a dimension originating from a common place. This starting, or "0," dimension is usually a datum. Holes are located to their centers. Other features are located to their edges. Refer to Example 1-17.

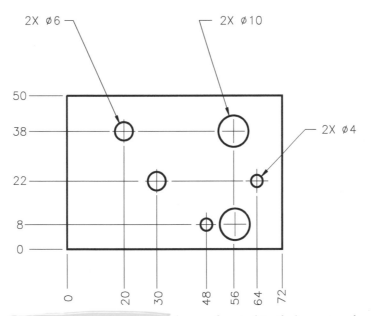

Example 1-17. Arrowless dimensioning. Holes are located to their centers from a datum. Other features are located to their edges from a datum.

Arrowless Tabular Dimensioning

Arrowless tabular dimensioning is the same as the arrowless dimensioning shown in Example 1-17, *except* the size dimensions that point to the holes with leader lines are omitted. In arrowless tabular dimensioning, identification letters are placed by the holes and the related information is given in a table. Refer to Example 1-18.

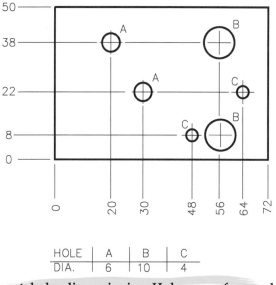

HOLE	A	B	C
DIA.	6	10	4

Example 1-18. Arrowless tabular dimensioning. Holes are referenced by letters. The related information is presented in a table.

Some companies take this practice one step farther and display the location and size of features in the table from an X and a Y axis. The depth of features is also provided from the Z axis where appropriate. Each feature is labeled with a letter or number that correlates to the table as shown in Example 1-19.

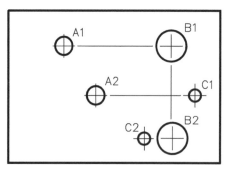

HOLE	QTY	DIA	X	Y	Z
A1	1	6	20	38	THRU
A2	1	6	30	22	THRU
B1	1	10	56	38	THRU
B2	1	10	56	8	THRU
C1	1	4	64	22	THRU
C2	1	4	48	8	THRU

Example 1-19. Holes may be located with X, Y, and Z references given in a table with arrowless tabular dimensioning.

Chart Dimensioning

Chart dimensioning may use dimension lines, arrowless, or arrowless tabular dimensions. This method provides flexibility in situations where dimensions change depending on the requirements of the product. The views of the part are drawn and variable dimensions are labeled with letters. These letters correlate with a chart where the different options are shown. On some parts there may be only one variable dimension, while others may contain several. Refer to Example 1-20.

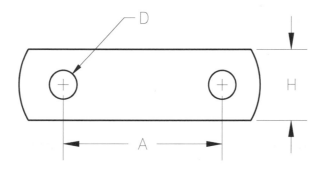

CHAIN NO.	A	D	H
SST1000	2.6	.44	1.125
SST1001	3.0	.48	1.525
SST1002	3.0	.95	2.125

Example 1-20. Chart dimensioning. Some parts may have many variable dimensions (as shown here). Other parts may have only one variable dimension.

GEOMETRIC DIMENSIONING AND TOLERANCING FOR CADD/CAM

The implementation of geometric tolerancing and dimensioning into a mechanical drafting CADD program is practical. The geometric tolerancing (GT) symbology makes this application a bonus to the mechanical drafting system. GT/CADD symbol libraries will be introduced in Chapter 2.

Some dimensioning and tolerancing guidelines for use in conjunction with CADD/CAM are outlined as follows:
❑ Geometric tolerancing is necessary to control specific geometric form and location.
❑ Major features of the part should be used to establish the basic coordinate system, but are not necessarily defined as datums.
❑ Subcoordinated systems that are related to the major coordinates are used to locate and orient features on a part.
❑ Define part features in relation to three mutually perpendicular reference planes, and along features that are parallel to the motion of CAM equipment.
❑ Establish datums related to the function of the part, and relate datum features in order of precedence as a basis for CAM usage.
❑ Completely and accurately dimension geometric shapes. Regular geometric shapes may be defined by mathematical formulas. A profile feature that is defined with mathematical formulas should not have coordinate dimensions unless required for inspection or reference.

❏ Coordinate or tabular dimensions should be used to identify approximate dimensions on an arbitrary profile.

❏ Use the same type of coordinate dimensioning system on the entire drawing.

❏ Continuity of profile is necessary for CADD. Clearly define contour changes at the change or point of tangency. Define at least four points along an irregular profile.

❏ Circular hole patterns may be defined with polar coordinate dimensioning.

❏ When possible, dimension angles in degrees and decimal parts of degrees. For example, 45°30′ should be dimensioned as 45.5°.

❏ Base dimensions at the mean of a tolerance because the computer numerical control (CNC) part programmer normally splits a tolerance and works to the mean. While this is theoretically desirable, one cannot predict where the part will be made. Dimensions should always be based on design requirements. If it is known that a part will be produced always by CNC methods, then establish dimensions without limits that conform to the CNC machine capabilities. Bilateral profile tolerances are also recommended for the same reason.

Test
Dimensioning and
Tolerancing

1

Name: _____

1. A(n) _____DIMENSION_____ is a numerical value expressed in appropriate units of measure, indicated on a drawing and in documents to define the size and/or geometric characteristics and/or locations of features of a part.

2. _____FEATURE_____ is a general term applied to a physical portion of a part.

3. Define tolerance. _____TOTAL AMOUNT A SPECIFIC_____ _____DIMENSION IS ALLOWED TO VARY_____ _____ _____

4. All dimensions shall have a tolerance except for dimensions that are identified as:
   a) reference.
   b) maximum.
   c) minimum.
   d) stock sizes.
   e) all of the above.

5. What are the limits of the dimension: 25±0.4? _____24.6   25.4_____

6. What is the tolerance of the dimension in question 5? _____.8_____

7. What is the specified dimension of the dimension shown in question 5? _____25_____

8. Give an example of an equal bilateral tolerance. _____.625 ±.005_____

9. Give an example of an unequal bilateral tolerance. _____.625 +.005 −.003_____

10. Give an example of a unilateral tolerance. _____.750 +.000 −.005_____

11. Define Maximum Material Condition (MMC). _____ _____CONDITION WHERE A FEATURE CONTAINS THE maximum_____ _____AMOUNT OF MATERIAL ALLOWED W/IN STATED LIMITS_____ _____

12. What is the MMC of the feature shown below? _____15.25_____

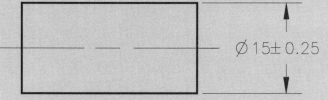

Ø15± 0.25

13. What is the MMC of the feature shown below? ___14.75___

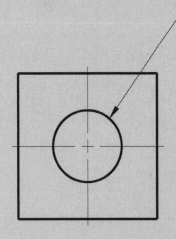

Ø15± 0.25

14. Define Least Material Condition (LMC). _____

___CONDITION WHERE A FEATURE CONTAINS THE LEAST___

___AMOUNT OF MATERIAL WITHIN ALLOWED LIMITS___

_____

_____

15. What is the LMC of the feature shown in question 12? ___14.75___

16. What is the LMC of the feature shown in question 13? ___15.25___

17. List the three general groups related to the standard ANSI fits between mating parts.

1) ___RUNNING & SLIDING___

2) ___FORCE___

3) ___LOCATIONAL___

18. Is the fit between the two parts shown below a clearance or a force fit?
    ___CLEARANCE___

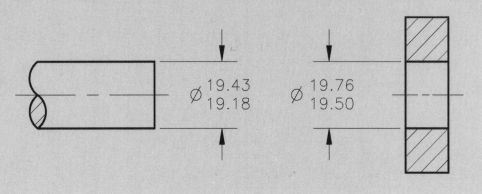

Ø 19.43 / 19.18   Ø 19.76 / 19.50

19. What is the allowance between the two parts shown in question 18?
    Show your calculations and label each numeral. _____

    | 19.50 | MMC HOLE |
    | 19.43 | mme SHAFT |
    | .07 | ALLOW |

20. What is the clearance between the two parts shown in question 18?
    Show your calculations and label each numeral. _____

    | 19.76 | LMC HOLE |
    | 19.18 | LMC SHAT |
    | .58 | CLEARANCE |

21. A force fit is also referred to as a(n) _____ SHRINK _____ or a(n)
    _____ fit.

22. Given the following information regarding the dimensions of a shaft and a collar (hole), determine the limits of the dimensions for each part. Show your calculations and label each numeral. Suggestion: review allowance and tolerance before you begin.

    a) The dimension of the shaft is $\emptyset14\pm0.4$.

    b) A clearance fit exists between the two parts.

    c) Provide an allowance 0.2.

    d) The tolerance to be applied to the collar hole dimension is 0.8.

    SHAFT UPPER LIMIT = ____ 14.4 ____

    SHAFT LOWER LIMIT = ____ 13.6 ____

    SHAFT TOLERANCE = ____ .8 ____

    | | MMC HOLE (UNKNOWN) | 14.6 |
    | − | MMC SHAFT | 14.4 |
    | = | ALLOWANCE = | .2 |
    | | MMC SHAFT | 14.4 |
    | + | ALLOWANCE | .2 |
    | = | MMC HOLE = | 14.6 |
    | | MMC HOLE | 14.6 |
    | + | HOLE TOLERANCE | .8 |
    | = | LMC HOLE = | 15.4 |
    | | HOLE LIMITS = | 14.6 / 15.4 |

23. Identify the ASME standard that is titled *Dimensioning and Tolerancing.* _____

    ASME Y14.5M

24. What does the abbreviation SI mean? _____

    INTERNATIONAL SYSTEM OF UNITS

25. What are the commonly used SI units on an engineering drawing? _____

    MILLIMETER

26. What are the commonly used US units found on engineering drawings? _____
    _____ INCHES _____

27. What general note should accompany a drawing to describe the predominant
    units used? ___UNLESS OTHERWISE SPECIFIED, ALL DIMENSIONS____
    ___ARE IN INCHES._____

28. Name the type of dimension that is placed on a drawing in parenthesis. _____
    _____

29. Identify the term that refers to the measured size of a feature or part after manu-
    facturing. ___FINISHED SIZE_____
    _____

30. Circle the letter of each of the following correct statements:

    (a) The decimal point and zero after it are omitted when the metric dimension is
        a whole number.

    (b) The last digit to the right of the decimal point is not followed by a zero when
        metric values are not equal to a whole number, unless needed to match the
        number of decimal places of the tolerance values.

    (c) A zero precedes a decimal metric dimension less than one millimeter.

    (d) Both the + and the – values of an inch tolerance have the same number of
        decimal places. Zeros are added to fill in where needed.

    e) A zero precedes a decimal inch value less than one.

    (f) A zero does *not* precede a decimal inch value less than one.

    (g) A specified dimension in inches is expressed to the same number of decimal
        places as its tolerance. Zeros are added to the right of the decimal point if
        needed.

    h) The manufacturing process, such as "DRILL" or "REAM," should accompany
        a dimension.

    i) Unless otherwise specified, all geometric tolerances apply for the full depth,
        length, and width of a feature.

31. Define "diameter." ___DISTANCE ACROSS A CIRCLE MEASURED__
    ___THRU THE CENTER_____
    _____
    _____
    _____

32. Define "radius." ___DISTANCE FROM THE CENTER OF A____
    ___CIRCLE TO THE OUTSIDE_____
    _____
    _____
    _____

33. Define "controlled radius." _A RADIUS THAT IS MAINTAINED_ _BETWEEN THE MAXIMUM & MINIUM LIMITS ALLOWED_ _AND HAS NO FEATURE REVERSALS ON THE CONTOUR._

34. True or False. Reversals in the contour are permitted with a radius dimension.

35. Define "Rule 1" in ASME Y14.5M–Extreme Form Variation. _THE FORM OF A FEATURE MAY VARY BETWEEN THE UPPER LIMIT & LOWER LIMIT OF A SIZE DIMENSION._

36. Define "limits of size." _BOUNDARY BETWEEN MMC & LMC_

Print Reading Exercise

1

Name:_____

The following print reading exercise uses actual industry prints with related questions that require you to read specific dimensioning and tolerancing representations. The answers should be based on the previously learned content of this book. The prints used are based on ASME standards. However, company standards may differ slightly. When reading these prints, or any other industry prints, a degree of flexibility may be required to determine how individual applications correlate with the ASME standards.

Print Reading Exercise

Refer to the print of the SLEEVE-DEWAR REIMAGING found on page 303.

1. Are the dimensions given in inches or millimeters?_____INCHES_____

2. What does the print say about burrs and sharp edges?_____FREE OF THEM_____

   _____

3. Is the part dimensioned using chain or datum dimensioning? _____DATUM_____

4. Refer to the 1.914 dimension:

   a) Where is the tolerance specified? _____TOL. BLOCK_____

   b) What is the tolerance?_____±.005_____

   c) What is the MMC?_____1.919_____

   d) What is the LMC? _____1.909_____

5. Refer to the ∅.8740±.0005 dimension:

   a) Where is the tolerance specified? _____@ THIS DIMENSION_____

   b) Is the tolerance unilateral or bilateral? _____BILATERAL_____

   c) What is the tolerance?_____±.0005_____

   d) What is the MMC?_____.8745_____

   e) What is the LMC? _____.8735_____

6. Refer to the ∅.750 dimension:

   a) Where is the tolerance specified? _____TOL. BLOCK_____

   b) What is the tolerance?_____+.003/−.001_____

   c) What is the MMC?_____.749_____

   d) What is the LMC? _____.753_____

7. Give the complete specifications associated with the ∅.107±.001 hole. _____.107" DIA ±.001" THRU, SPOTFACE .218" DIA +.003/−.001_____

   _____

8. What does the circle on the leader connected to the 2X 45°X.010 dimension mean? (Look at Example 2-1 in Chapter 2.) ___ALL AROUND___

_____

_____

Refer to the print of the BRACKET found on page 304.

9. Are the dimensions in inches or millimeters? ___INCHES___

10. Is datum or chain dimensioning used? ___DATUM___

11. Refer to the ∅.875±.005 dimension:

    a) What is the specified dimension? ___.875"___

    b) What is the tolerance? ___±.005" = .010"___

    c) What is the MMC? ___.870"___

    d) What is the LMC? ___.880"___

12. What does the box around the 1.950 dimension mean? (Look at Example 2-11 in Chapter 2.) ___BASIC DIMENSION___

13. What does the note "INTERPRET DIMENSIONS AND TOLERANCES PER ASME Y14.5M-1994" mean? _____

_____

_____

_____

14. What does ASME stand for? (Read the Introduction.) _AMER SOCIETY OF MECHANICAL ENGRS_

Refer to the print of the HUB-STATIONARY ATU found on page 305.

15. Are the dimensions in inches or millimeters? ___INCHES___

16. Refer to the ∅4.4997–4.4994 dimension:

    a) What is the tolerance? ___.0003"___

    b) What is the MMC? ___4.4997___

    c) What is the LMC? ___4.4994___

17. Refer to the ∅3.900±.005 dimension:

    a) What is the specified dimension? ___3.900___

    b) What is the tolerance? ___.010___

    c) What is the MMC? ___3.895___

    d) What is the LMC? ___3.905___

18. Refer to the ∅.352+.005/−.001 dimension:

    a) How many of these features are there? ___6___

    b) What are the limits of the dimension? ___.357 / .351___

    c) What is the MMC? ___.351___

    d) What is the LMC? ___.357___

Refer to the print of the PEDAL-ACCELERATOR found on page 306.

19.   Are the dimensions given in inches or millimeters?___MILLIMETERS___

20.   Refer to the 2X 45°X0.76 dimension:

    a)   What is the tolerance on the 0.76? ___1.0___

    b)   What is the tolerance on the 45°? ___10°___

21.   Refer to the Ø4.834–4.763 dimension:

    a)   What is the MMC?___4.763___

    b)   What is the LMC? ___4.834___

    c)   What is the tolerance?___.071___

Refer to the print of the MOUNTING PLATE (UPPER)-FRAME ASSY 3 AXIS HP found on page 307.

22.   What do the parentheses around the 1.875 dimension mean? ___
___REFERENCE DIMENSION___

23.   Refer to the 1.60 dimension:

    a)   Where is the tolerance found?___TOLERANCE BLOCK___

    b)   What is the tolerance?___.030___

    c)   What are the limits? ___1.615   1.585___

    d)   What is the MMC?___1.615___

    e)   What is the LMC? ___1.585___

Refer to the print of the HYDRAULIC VALVE found on page 308.

24.   Refer to the Ø.961/.959 dimension:

    a)   What is the tolerance?___.002___

    b)   What is the MMC?___.961___

    c)   What is the LMC? ___.959___

25.   What is dimension "A" for part number 1 MS 2427-2?   Ø.715/.718

    a)   What is the MMC of this dimension? ___.715___

    b)   What is the LMC of this dimension? ___.718___

    c)   What is the tolerance for this dimension?___.003___

26.   Name and describe the dimensioning practice used for the dimension described in question number 25. ___TABULATED OR CHART DIMENSIONS___
___
___
___
___

27.   What is the tolerance for the 45° dimension?___4°___

Refer to the print of the COVER, CAGE-INNER AZ DRIVE found on page 309.

28. Refer to the SR7.500 dimension:

   a) What does the "SR" mean? _____ SPHERICAL RADIUS _____

   b) What is the tolerance? _____ .020 _____

   c) What is the MMC? _____ 7.510 _____

   d) What is the LMC? _____ 7.490 _____

Refer to the print of the HOUSING-LENS, FOCUS found on page 313.

29. Name and describe the type of dimensioning practice found on this print. _____
   _____ ARROW LESS DIMENSIONING _____
   _____
   _____
   _____

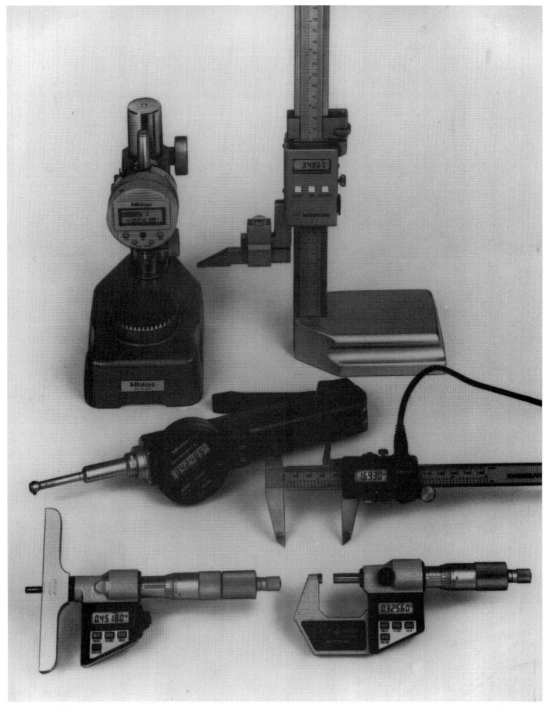

Many different measuring devices are used in the inspection of parts manufactured to meet geometric dimensioning and tolerancing specifications. (Mitutoyo)

Chapter 2

An Introduction to Symbols, Terms, and Computer-aided Design and Drafting

This chapter helps you in the identification of symbols and terms. Your main objective is to recognize the various types of symbols by their name, shape, and size. Only a few terms are defined at this time. Other terms will be clearly defined in later chapters as you learn about geometric tolerancing. Symbol sizes are based on drawing lettering height. Where symbols are detailed you will see the note "h = lettering height." This means that "h" equals the predominant lettering height on the drawing. For example, the lettering height on most engineering drawings is .125 inch or 3 mm, depending on company standards. Verify standard lettering height with ASME Y14.2M-1992 *Line Conventions and Lettering.*

Dimensioning and geometric tolerancing symbols are divided into five basic types:

1. Dimensioning symbols.

2. Datum feature and datum target symbols.

3. Geometric characteristic symbols.

4. Material condition symbols.

5. Feature control frame.

When you draw symbols on test answers or in problem solutions, use clear, accurate representations. It is also recommended that an appropriate geometric tolerancing template be used for manual drafting, or a symbol library for computer-aided design and drafting (CADD). Geometric tolerancing symbols are drawn using thin lines that are the same thickness as extension and dimension lines (.01 inch or 0.3 mm).

DIMENSIONING SYMBOLS

Symbols represent specific information that would otherwise be difficult and time consuming to duplicate in note form. Symbols must be clearly drawn to the required size and shape so they communicate the desired information uniformly. Symbols are recommended by ASME Y14.5M because symbols are an international language and are read the same way in any country. It is important in an international economy to have effective communication on engineering drawings. Symbols make this communication process uniform. ASME Y14.5M also states that the adoption of dimensioning symbols does not prevent the use of equivalent terms or abbreviations in situations where symbols are considered inappropriate.

Symbols aid in clarity, ease of drawing presentation, and save time, especially when used in conjunction with computer-aided design and drafting (CADD). Symbols should be drawn clearly using a template or CADD. Creating and using CADD symbols is covered later in this chapter. Example 2-1 shows recommended dimensioning symbols.

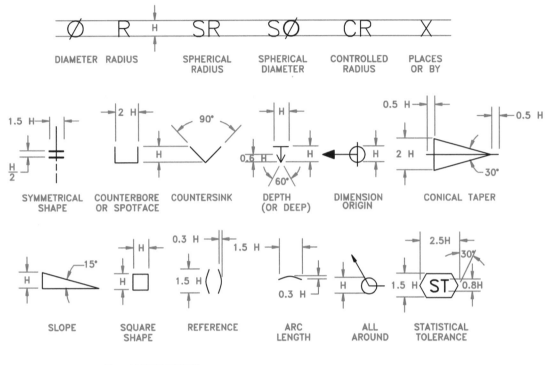

Example 2-1. Recommended ASME dimensioning symbols.

DIMENSIONING AND TOLERANCING TEMPLATES

Dimensioning and tolerancing templates are available to help save time when doing drawings that contain ASME Y14.5M symbols. It is recommended that one of these templates be used when doing the exercises or tests in this book. The symbols on drawing assignments should be done using a template or a CADD system. A template or CADD presentation helps ensure that all symbols are properly drawn and always uniform in size and appearance. Standard lettering height on engineering mechanical drawings is .125 inch or 3 mm, depending on company standards. The symbols are sized based on the predominant lettering height on the drawing. Verify the lettering height specified when purchasing a template. The template shown in Example 2-2 has both geometric tolerancing and standard dimensioning symbols.

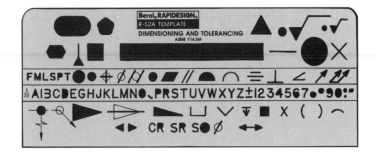

Example 2-2. A geometric dimensioning and tolerancing template.

DATUM FEATURE SYMBOL

Datums are considered theoretically perfect surfaces, planes, points, or axes. This is only an introduction to the appearance of datum-related symbols. Chapter 3 provides a complete discussion on datums and datum identification. As you saw in Chapter 1, datum dimensioning is where all dimensions originate from a common surface (refer to Example 1-16). This is also true where a group of dimensions originate from a common axis or center plane. In these applications the datum is assumed. In geometric dimensioning and tolerancing, the datums are identified with a *datum feature symbol.*

Any letter of the alphabet may be used to identify a datum except for I, O, or Q. These letters may be confused with the numbers 1 or 0. Each datum feature requiring identification must have its own identification letter. On drawings where the number of datums exceed the letters in the alphabet, then double letters are used starting with AA through AZ, and then BA through BZ. Datum feature symbols may be repeated only as necessary for clarity. Datum identification letters A, B, and C may be used for convenience; however, other letters are commonly used in industry. Example 2-3 shows a datum feature symbol.

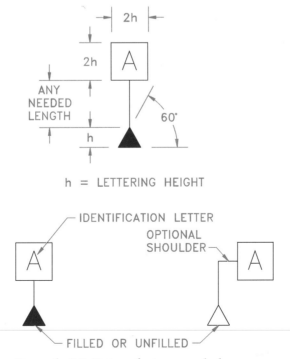

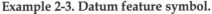

Example 2-3. Datum feature symbol.

DATUM TARGET SYMBOLS

Datum targets are used to specify points, lines, or areas of contact on a part that establish datums when it is not possible to use a surface. This discussion is provided for you to see what the symbols related to datum targets are supposed to look like. Further explanation is provided in Chapter 3.

The *datum target symbol* is drawn as a circle with a horizontal line through the center. The top half of the circle is left blank unless the datum target symbol refers to a datum target area. In that case, the size of the target area is specified inside the symbol, as shown in Example 2-4. A dot at the end of a leader pointing to the inside of the symbol may also be used to specify a datum target area when there is not enough room for the note to be placed inside. The lower half of the circle is used to identify the related datum with the datum reference letter and datum target number assigned sequentially, starting with 1, for each datum, such as A1, A2, and A3. Refer to Example 2-4.

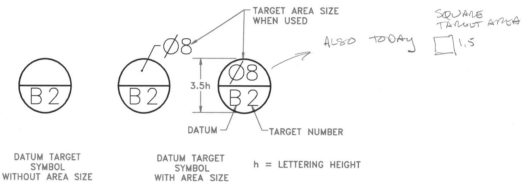

DATUM TARGET
SYMBOL
WITHOUT AREA SIZE

DATUM TARGET
SYMBOL
WITH AREA SIZE

h = LETTERING HEIGHT

Example 2-4. Datum target symbol.

A radial line is used to connect the datum target symbol to the datum target point, target line, or target area, shown in Example 2-5. These three examples of datum target symbols are applied on datums in Chapter 3.

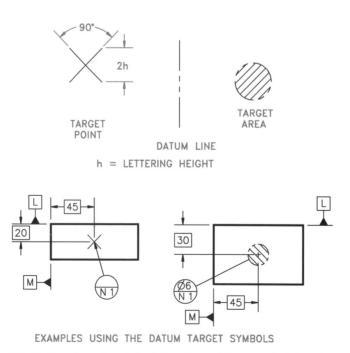

EXAMPLES USING THE DATUM TARGET SYMBOLS

Example 2-5. Datum target point, datum target line, and datum target area.

GEOMETRIC CHARACTERISTIC SYMBOLS

Symbols used in geometric dimensioning and tolerancing to provide specific controls related to the form of an object, the orientation of features, the outlines of features, the relationship of features to an axis, or the location of features are known as geometric characteristic symbols. *Geometric characteristic symbols* are separated into five types: form, profile, orientation, location, and runout, as shown in Example 2-6.

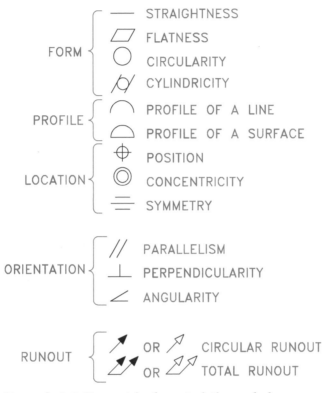

Example 2-6. Geometric characteristic symbols.

The symbols in Example 2-6 are drawn to the actual size and shape recommended by ASME Y14.5M based on .125 inch high lettering.

MATERIAL CONDITION SYMBOLS

Material condition symbols are often referred to as "modifying symbols" because they modify or change the geometric tolerance in relation to the produced size or location of the feature. Material condition symbols are only used in geometric dimensioning applications. The symbols used in the feature control frame to indicate maximum material condition (MMC) or least material condition (LMC) are shown in Example 2-7. Regardless of feature size (RFS) is also a material condition. However, there is no symbol for RFS because it is assumed for all geometric tolerances and datum references unless MMC or LMC is specified.

This is only an introduction to material condition symbols. Chapter 4 deals exclusively with this subject.

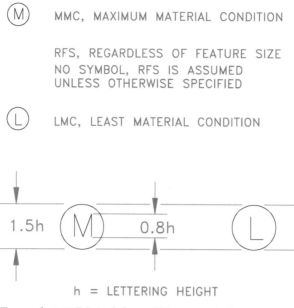

Example 2-7. Material condition symbols.

FEATURE CONTROL FRAME

A geometric characteristic, geometric tolerance, material condition, and datum reference (if any) for an individual feature is specified by means of a feature control frame. The *feature control frame* is divided into compartments containing the geometric characteristic symbol in the first compartment followed by the geometric tolerance. Where applicable, the geometric tolerance is preceded by the diameter symbol that describes the shape of the tolerance zone and followed by a material condition symbol if other than RFS. See Example 2-8.

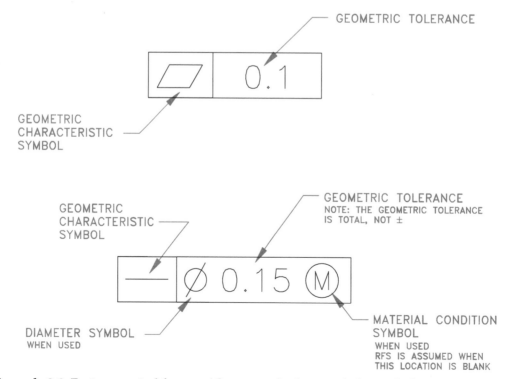

Example 2-8. Feature control frame with geometric characteristic symbol, geometric tolerance, diameter symbol (when used), and material condition symbol (when used).

Where a geometric tolerance is related to one or more datums, the datum reference letters are placed in compartments following the geometric tolerance. Where a datum reference is multiple (that is, established by two datum features such as an axis established by two datum diameters) both datum reference letters, separated by a dash, are placed in a single compartment after the geometric tolerance. This is known as *multiple datum reference.* Example 2-9 shows several feature control frames with datum references.

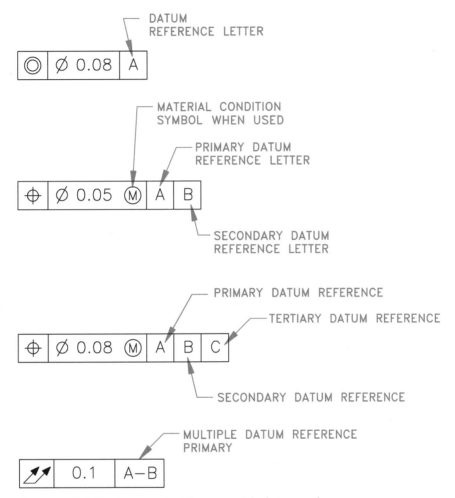

Example 2-9. Feature control frames with datum references.

The order of elements in a feature control frame is shown in Example 2-10. Notice in Example 2-10 that the datum reference letters may be followed by a material condition symbol where applicable. Draw each feature control frame compartment large enough to accommodate the symbols without crowding. Minimum compartment length is 2 × the lettering height. Maintain the minimum compartment sizes when the symbols or letters fit without crowding.

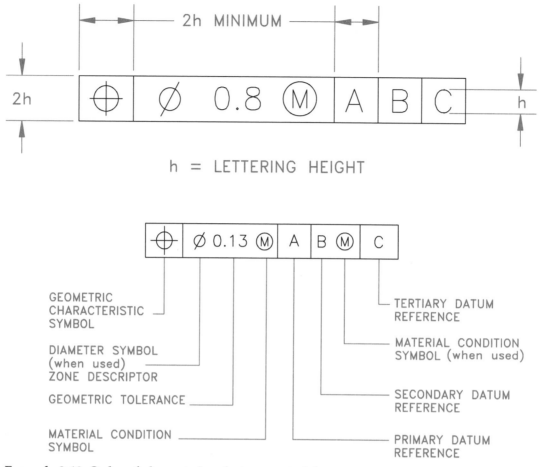

Example 2-10. Order of elements in a feature control frame.

BASIC DIMENSIONS

A *basic dimension* is considered a theoretically perfect dimension. Basic dimensions are used to describe the theoretically exact size, profile, orientation, or location of a feature or datum target. These dimensions provide the basis where permissible variations are established by tolerances on other dimensions, in notes, or in feature control frames. In simple terms, all a basic dimension does is tell you where the geometric tolerance zone or datum target is located. This text will show you specific situations where basic dimensions are optional or required. Basic dimensions are shown on a drawing by placing a rectangle around the dimension as shown in Example 2-11. A general note may also be used to identify basic dimensions in some applications. For example, the note "UNTOLERANCED DIMENSIONS LOCATING TRUE POSITION ARE BASIC" indicates the dimensions that are basic.

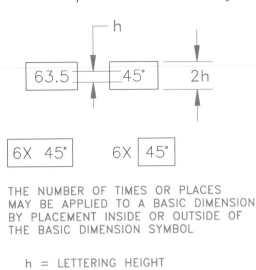

Example 2-11. Basic dimensions.

The basic dimension symbol around a dimension is a signal to the reader to look for a geometric tolerance in a feature control frame related to the features being dimensioned.

Drawing Basic Dimensions with CADD

AutoCAD® allows you to automatically draw basic dimensions by checking the "Basic Dimension" box in the "Dimension Line" dialog box. Access the "Dimension Line" dialog box by picking the "Dimension Line..." button in the "Dimension Styles and Variables" dialog box. This is accessed by picking "Dimension Style..." from the "Settings" pull-down menu or by typing "DDIM" at the "Command:" prompt. A basic dimension drawn using AutoCAD® is shown in Example 2-12.

Example 2-12. A basic dimension drawn using AutoCAD®.

ADDITIONAL SYMBOLS

There are a few other symbols used in geometric dimensioning and tolerancing. These symbols are used for specific applications and are identified as follows (also refer to Example 2-13):

Free state describes distortion of a part after the removal of forces applied during manufacture. The free state symbol is placed in the feature control frame after the geometric tolerance and the material condition (if any) if the feature must meet the tolerance specified while in free state. See Chapter 5 for more detail on this subject.

A *tangent plane* symbol is placed after the geometric tolerance in the feature control frame when it is necessary to control a feature surface by contacting points of tangency. See Chapter 6 for more detail on this subject.

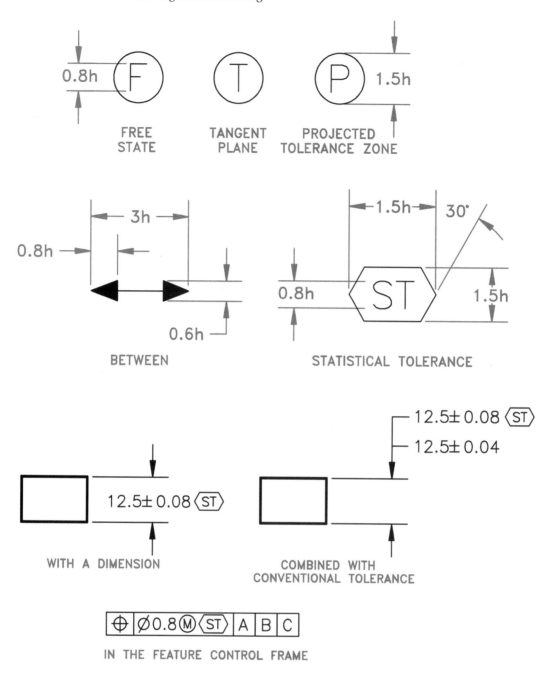

STATISTICAL TOLERANCING METHODS

Example 2-13. Additional dimensioning symbols.

A *projected tolerance zone* symbol is placed in the feature control frame to inform the reader that the geometric tolerance zone is projected away from the primary datum. See Chapter 7 for more detail on this subject.

The *between* symbol is used with profile geometric tolerances to identify where the profile tolerance is applied. See Chapter 5 for more detail on this subject.

The *statistical tolerance* symbol is used to indicate that a tolerance is based on statistical tolerancing. *Statistical tolerancing* is the assigning of tolerances to related dimensions in an assembly based on the requirements of statistical process control (SPC). *Statistical process control* is a method of monitoring a manufacturing process by using statistical signals to either leave the process alone or change it as needed to maintain the quality intended in the dimensional tolerancing. The statistical tolerancing symbol is placed after the dimension or geometric tolerance that requires SPC. When the feature may be manufactured either by using SPC or by using conventional means, both the statistical tolerance with the statistical tolerance symbol and the conventional tolerance must be shown. See Example 2-13. An appropriate general note should accompany the drawing. Either of the two notes shown below are acceptable:

❏ FEATURES IDENTIFIED AS STATISTICAL TOLERANCED SHALL BE PRODUCED WITH STATISTICAL PROCESS CONTROL.

❏ FEATURES IDENTIFIED AS STATISTICAL TOLERANCED SHALL BE PRODUCED WITH STATISTICAL PROCESS CONTROL, OR THE MORE RESTRICTIVE ARITHMETIC LIMITS.

A complete display of all dimensioning and geometric tolerancing symbols is provided in Appendix A5 for your convenience.

COMPUTER-AIDED DESIGN AND DRAFTING (CADD)

One of the advantages of using computer-aided design and drafting (CADD) is the potential increase in productivity over manual drafting methods. This increase in productivity is achieved when the standard symbols used in drafting are created in a menu. The menu is a list of items such as computer commands, numbers, or symbols that the drafter can select as needed. There are several types of menus including keyboard menus, screen menus, tablet menus, button menus, and auxiliary menus.

The typewriter keyboard with additional function keys is commonly used with CADD workstations. Alphanumeric data (letters and numbers) may be entered into the computer to implement commands and drawing information.

The screen menu contains commands that may be selected using function keys, puck keys, a stylus, a mouse, or by touch, depending on how the system is set up.

A tablet or digitizer menu is set up with commands and symbols that are customized for a specific purpose. This customized menu is often referred to as a symbol library, symbol directory, cell library, or CADD template.

Puck or mouse keys are used as button menus where screen or digitizer commands are selected by pressing a specific key (button). Additional buttons may be programmed to initiate a variety of commonly used commands.

Auxiliary menus include voice activated and auxiliary keyboard menus. Auxiliary keyboards are similar to the use of puck keys. Computer functions and commands are activated by selecting the specific key or button on the auxiliary keyboard.

The advantage of using CADD exists because geometric dimensioning and tolerancing symbols can be placed on a drawing in much less time than it takes to manually draw the same symbol. The added advantage occurs when it is necessary to revise or change a drawing. The process of making engineering changes that often take many hours manually may be done in a few minutes using CADD. It becomes a simple task to either remove or change symbols on the computer and then print or plot a new representation.

Pointing Devices

The most popular method of selecting items on a menu tablet is with a pointing device. *Pointing devices* refer to a variety of instruments that are attached to the digitizer or computer terminal. Some systems allow the drafter's finger to be used as the pointing device. These systems have function boards, buttons, touch-sensitized menu boards, or touch-sensitive computer screens.

A *digitizer cursor* is the most common pointing device. A digitizer cursor is an input device that is held in the hand. The screen cursor is a small box or lines crossing on the monitor screen that indicates the current position. The puck and stylus shown in Example 2-14 are two forms of digitizer cursors.

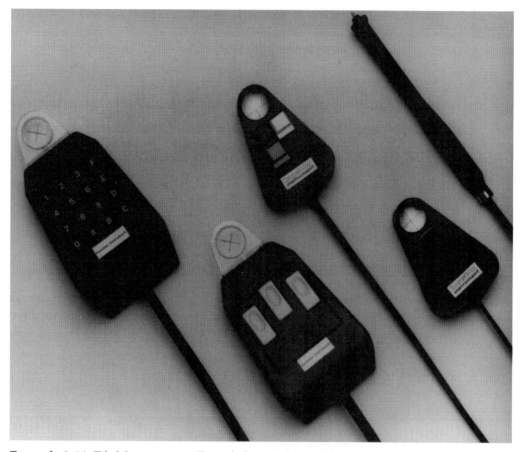

Example 2-14. Digitizer cursors. From left to right: a 16-button puck, a 3-button puck, a 4-button puck, a 1-button puck, and a stylus. (Courtesy of Houston Instruments, A Summagraphics Company.)

A *light pen* is another type of pointing device. A light pen allows a drafter to digitize information into the computer by pressing the "pen" on the screen at the desired location.

A *mouse* is a device that senses its position on a flat surface by movement of a ball or by reflected light across a grid. When the mouse is moved across a flat surface, the screen cursor also moves. Buttons on the mouse activate specific functions or allow the user to choose from menu items displayed on the screen. Refer to Example 2-15.

Example 2-15. A mouse can be used to select items from a screen menu.

A *trackball* can be described as an "up-side-down" mouse. The user rotates a ball that is located on the top of the unit. That movement is in turn translated by the computer into cursor movement. The whole trackball unit stays stationary, whereas with a mouse the unit moves.

Digitizer

A *digitizer* is also referred to as a *menu tablet* or a *graphics tablet.* A digitizer is an electronic input device that allows data to be entered into the computer by pointing using a pointing device. The puck or stylus senses movement through a magnetic field in the digitizer tablet. Several digitizer sizes are shown in Example 2-16.

Example 2-16. A variety of digitizer sizes. (Courtesy of Houston Instruments, A Summagraphics Company.)

When the pointing device is placed on the digitizer cursor crosshairs or an aperture box, the video display screen registers the location. In most situations, a printed menu (referred to as a *digitizer tablet menu*) is placed on the digitizer tablet. Commands are displayed on the tablet menu that are used for specific applications such as mechanical or architectural drafting. Menus may also be customized with special symbols such as geometric tolerancing. A sample standard CADD menu, referred to as a *template,* is shown in Example 2-17. Notice that the top portion of this menu has been customized with GD&T symbols.

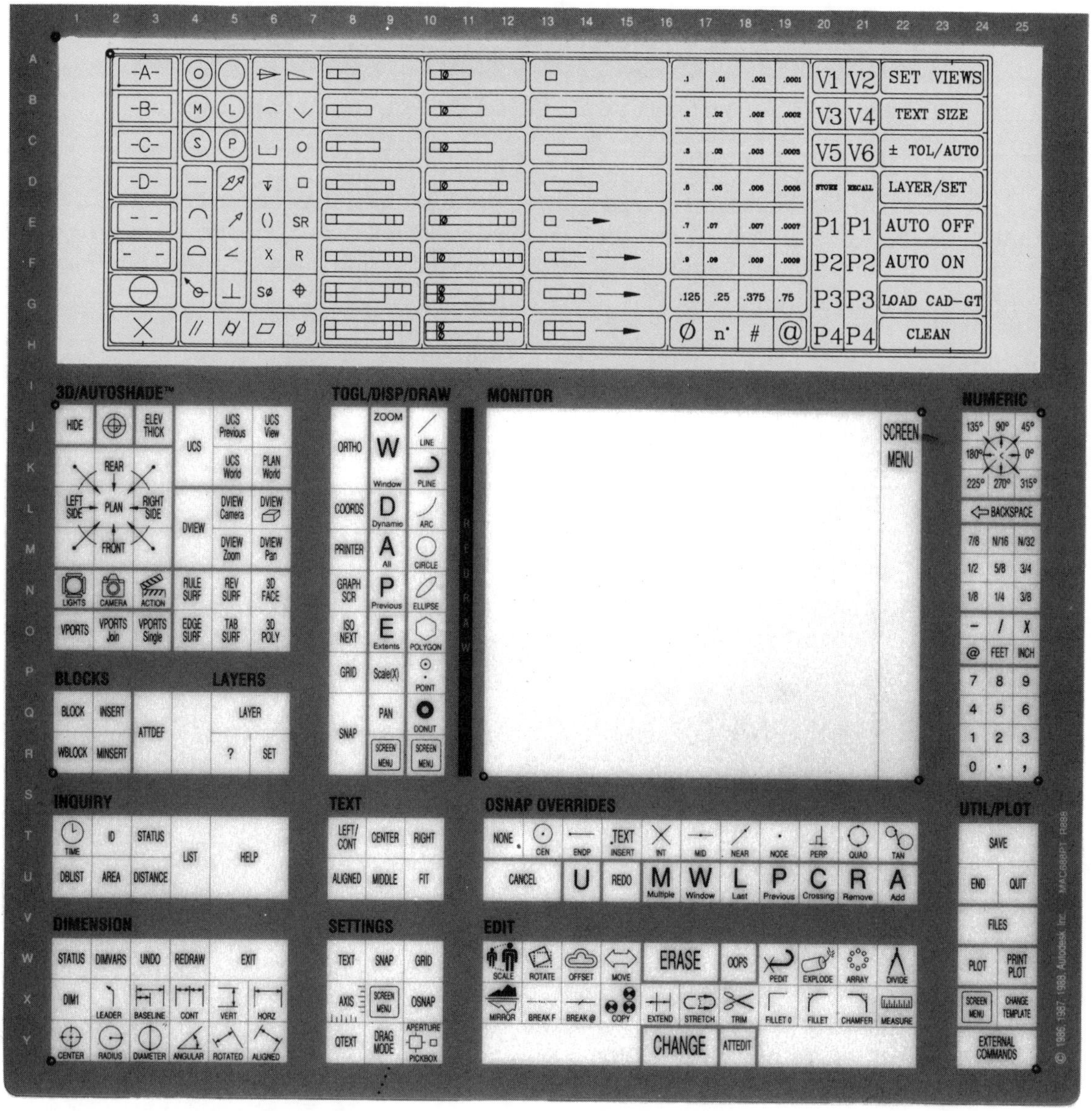

Example 2-17. A standard CADD menu. The top portion of this template may be customized with special symbol libraries such as the GD&T library shown here. (Courtesy of Autodesk, Inc.)

The easiest way to use the menus is to place registration tabs on the digitizer that align with the prepunched holes on the preprinted menu sheet.

Menus are often made of polyester so they are durable and remain dimensionally stable. Customized overlays of specific symbols may be registered over the standard menu. The advantage of this is that several custom symbol libraries or templates can be designed and easily substituted by removing one and placing the next over the standard menu. An advantage of the tablet menu is the user can pick commands by association with a symbol on the menu.

CADD GENERATED GEOMETRIC TOLERANCING SYMBOLS

Before you begin creating a geometric tolerancing symbol, refer to this text to determine the proper size and format for the ASME Y14.5M symbols. Make a sketch of the desired symbol and decide how it will be placed on a drawing. Name the symbol–this will also be the CADD file name. Select a point on each symbol that is a convenient point of origin for placing the symbol on a drawing. The point of origin is the location position for placing a symbol on a drawing. Example 2-18 shows the insertion point determined for a sample of geometric tolerancing symbols. The insertion point may be any corner of the symbol, but you should consider the insertion point as the most common point of origin.

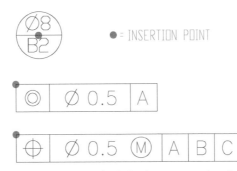

Example 2-18. Selecting the point of origin for geometric tolerancing symbols.

After you have drawn a specific symbol, it must be stored as a symbol by using commands such as the AutoCAD® BLOCK or WBLOCK commands. Refer to *AutoCAD and its Applications* or *AutoCAD and its Applications for Windows* published by The Goodheart-Willcox Company, Inc., and see the chapter entitled "Customizing Auto-CAD's Tablet Menu" for a complete discussion on this subject for AutoCAD® applications. When all of the symbols have been placed on the template, assign a name such as "GEOMETRIC TOLERANCING." Print out or plot the symbols on a sheet so they correspond to the locations where they were placed on the template. This new symbol library can then be placed as an overlay on the menu tablet.

GEOMETRIC DIMENSIONING AND TOLERANCING WITH AUTOCAD® RELEASE 13

The previous section gave a brief introduction to the use of customized tablet menus and symbol libraries for CADD geometric dimensioning and tolerancing applications. AutoCAD® Release 13 has the ability to add GD&T symbols to your drawings. The feature control frame and related GD&T symbols can be created using the tolerance and leader commands.

Using the Tolerance Command

The tolerance command provides tools for creating GD&T symbols. You can access this command in one of three ways.

❏ Pick **Toleran:** from the **DRAW DIM** screen menu.
❏ Pick **Dimensioning** followed by **Tolerance**... in the **Draw** pull-down menu.
❏ Type either **TOL** or **TOLERANCE** at the **Command:** prompt.

When you enter any of these commands, the **Symbol** dialog box shown in Example 2-19 appears. This dialog box contains the geometric characteristic symbols. When you pick a desired symbol, it becomes highlighted. The last option is blank. Next, pick the **OK** button. The **Geometric Tolerance** dialog box shown in Example 2-20 appears. The symbol that you picked in the **Symbol** dialog box is displayed in the **Sym** text box. The **Geometric Tolerance** dialog box is divided into compartments that relate a feature control frame.

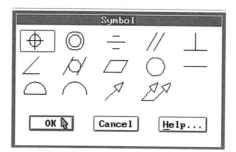

Example 2-19. The **Symbol** dialog box is where you select the symbol you need.

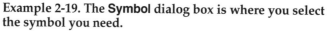

Example 2-20. The **Geometric Tolerance** dialog box is used to build a feature control frame.

The **Tolerance 1** compartment allows you to enter the first geometric tolerance value found in the feature control frame. If you pick below **Dia** and **MC,** you can add diameter and material condition symbols.

There are two text edit boxes in this compartment. The first text edit box is for the information found in a single feature control frame. The second text edit box is for information needed for a double feature control frame. Double feature control frames are used for applications such as unit straightness, unit flatness, composite profile tolerance, composite positional tolerance, or coaxial positional tolerance.

The **Tolerance 2** compartment is used for the addition of a second geometric tolerance to the feature control frame. This is not a common application, but it may be used in some cases where there are restrictions placed on the geometric tolerance specified in the first compartment. For example, 0.8 MAX might be entered. This means that the specification given in the first compartment is maintained, but may not exceed 0.8 maximum (MAX) as given in the second compartment.

The **Datum 1** box is used to establish the information needed in the primary datum reference compartment. You can also enter a material condition symbol by picking **MC** if needed. The **Datum 2** and **Datum 3** boxes work the same as **Datum 1,** but are for setting the secondary and tertiary datum reference information. Datums are covered in more detail in Chapter 3.

The **Height** box allows you to enter the height of a projected tolerance zone. Pick to the right of **Projected Tolerance Zone** to access the symbol. Projected tolerance zones are discussed in Chapter 8.

In the **Datum Identifier** box, you can specify a datum feature symbol in this section for the ANSI Y14.5M-1982 standard. You need to design a datum feature symbol and save it as a block if you want to comply with ANSI/ASME Y14.5M-1994.

When you have entered all of the desired information in the **Geometric Tolerance** dialog box, pick the **OK** button. The following prompt appears.

> Enter tolerance location:

Pick the place for the feature control frame to be drawn. The feature control frame from the previous command sequence is shown in Example 2-21.

Using the Leader Command

You can also use the **LEADER** command to access the same dialog boxes and draw a feature control frame connected to a leader line. With this command, you first draw a leader and shoulder. Then use the **Annotations** and **Tolerance** options to access the **Symbol** dialog box. The rest of the procedure is the same as above.

Other Resources

For a comprehensive understanding of how to use AutoCAD®, refer to *AutoCAD and its Applications—Basics, Release 13 for DOS* or *AutoCAD and its Applications—Basics, Release 13 for Windows.* Advanced books are also available for both DOS and Windows. All of these books, and books covering previous releases of AutoCAD®, are available directly from publisher.

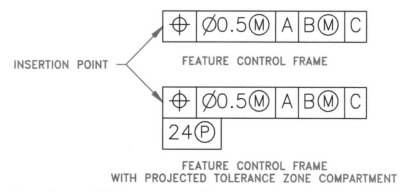

Example 2-21. This feature control frame was drawn using AutoCAD® Release 13.

Name: _____

A dimensioning and tolerancing template is recommended for drawing proper symbols on this test and on future tests.

1. List the five basic types of geometric dimensioning and tolerancing symbols.

    1) _____

    2) _____

    3) _____

    4) _____

    5) _____

2. Name the five types of geometric characteristic symbols.

    1) _____

    2) _____

    3) _____

    4) _____

    5) _____

3. Name each of the following geometric characteristic symbols.

    ___ _____   ◎ _____

    ▱ _____   ≡ _____

    ○ _____   // _____

    ⌀ _____   ⊥ _____

    ⌒ _____   ∠ _____

    ◠ _____   ↗ _____

    ⊕ _____   ↗↗ _____

4. Any letter of the alphabet can be used to identify a datum except for _____, _____, or _____.

5. When may datum feature symbols be repeated on a drawing? _____

    _____

    _____

6. What information is placed in the lower half of the datum target symbol?_____

    _____

    _____

7. What information is placed in the top half of the datum target symbol? _____

_____

_____

_____

8. Label the parts of the following feature control frame.

_____

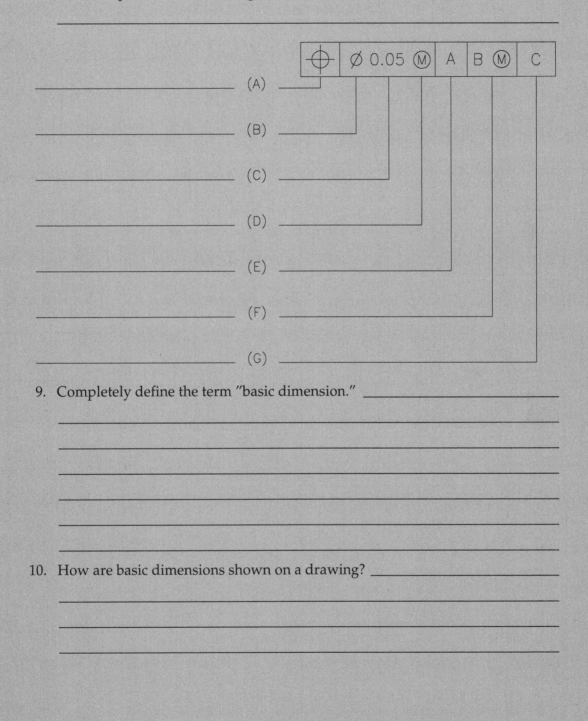

_____ (A)

_____ (B)

_____ (C)

_____ (D)

_____ (E)

_____ (F)

_____ (G)

9. Completely define the term "basic dimension." _____

_____

_____

_____

_____

_____

_____

10. How are basic dimensions shown on a drawing? _____

_____

_____

_____

11. Name the following symbols.

Print Reading Exercise

2

Name: _____

The following print reading exercise uses actual industry prints with related questions that require you to read specific dimensioning and geometric tolerancing representations. The answers should be based on previously learned content of this book. The prints used are based on ASME standards, however company standards may differ slightly. When reading these prints, or any other industry prints, a degree of flexibility may be required to determine how individual applications correlate with the ASME standards.

Print Reading Exercise

Refer to the print of the SLEEVE-DEWAR REIMAGING found on page 303.

1. Identify the dimensioning and tolerancing standard that was used as a basis for the dimensioning and tolerancing placed on this drawing. _____

    _____

    _____

2. List the names of at least six dimensioning symbols, excluding geometric tolerancing symbols, found on this print.

    a) _____

    b) _____

    c) _____

    d) _____

    e) _____

    f) _____

3. List the names of at least ten geometric dimensioning and tolerancing symbols found on this print.

    a) _____

    b) _____

    c) _____

    d) _____

    e) _____

    f) _____

    g) _____

    h) _____

    i) _____

    j) _____

4. Give a complete identification for each of the items found in the feature control frame associated with the ∅.750 dimension. Identify the items from left to right.

   a) _____

   b) _____

   c) _____

   d) _____

   e) _____

   f) _____

Refer to the print of the HUB-STATIONARY, ATU found on page 305.

5. List the names of at least five dimensioning symbols, excluding geometric tolerancing symbols, found on this print.

   a) _____

   b) _____

   c) _____

   d) _____

   e) _____

6. List the names of at least eight geometric dimensioning and tolerancing symbols found on this print.

   a) _____

   b) _____

   c) _____

   d) _____

   e) _____

   f) _____

   g) _____

   h) _____

7. Give a complete identification for each of the items found in the feature control frame associated with the ∅4.500 dimension. Identify these items from left to right.

   a) _____

   b) _____

   c) _____

   d) _____

8. Why is there no material condition symbol in the feature control frame described in question number 7? _____

_____

_____

_____

Refer to the print of the PEDAL-ACCELERATOR found on page 306.

9. List the names of at least nine geometric dimensioning and tolerancing symbols found on this print.

a) _____

b) _____

c) _____

d) _____

e) _____

f) _____

g) _____

h) _____

i) _____

Refer to the print of the HYDRAULIC VALVE found on page 308.

10. List the names of at least eight geometric dimensioning and tolerancing symbols found on this print.

a) _____

b) _____

c) _____

d) _____

e) _____

f) _____

g) _____

h) _____

Refer to the print of the PLATE-TOP MOUNTING found on page 311.

11. List the names of the different geometric characteristic symbols found on this print.

a) _____

b) _____

c) _____

d) _____

e) _____

f) _____

Refer to the print of the BRACKET ASSY-EL GIMBAL found on page 315.

12. List the names of at least eight dimensioning symbols, excluding geometric dimensioning and tolerancing symbols, found on this print.

    a) _____

    b) _____

    c) _____

    d) _____

    e) _____

    f) _____

    g) _____

    h) _____

13. List the names of at least five geometric dimensioning and tolerancing symbols found on this print.

    a) _____

    b) _____

    c) _____

    d) _____

    e) _____

Chapter 3

Datums

Datums are considered theoretically perfect planes, surfaces, points, lines, or axes. Datums are placed on drawings as requirements for referencing features of an object, as in datum dimensioning that was discussed in Chapter 1.

These datums are used by the machinist, toolmaker, or quality control inspector to ensure that the part is in agreement with the drawing.

This chapter is designed to help you identify and read information related to datums on drawings. This chapter also covers the specifications for properly placing datum related symbols on drawings. This information is covered without regard to specific inspection and tooling techniques. Advanced instruction is recommended after a solid understanding of the basic fundamentals presented here.

DATUMS

Datums are planes, surfaces, points, lines, or axes where measurements are made from. A datum is assumed to be exact. A *datum feature* is an actual feature on a part, such as a surface, that is used to establish a datum. A datum is the true geometric counterpart of a datum feature. Datums are placed on drawings as requirements for referencing features of an object. Examples of datums in manufacturing are machine tables, surface plates, gauge surfaces, surface tables, or specially designed rotation devices. These are referred to as *datum feature simulators* and are used to contact the datum features and establish what are known as the *simulated datums.*

Location and size dimensions are established from the datum. There are many concepts to keep in mind when datums are established, including the function of the part or feature, manufacturing processes, methods of inspection, the shape of the part, relationship to other features, assembly considerations, and design requirements. Datum features should be selected to match on mating parts, should be easily accessible, and should be of adequate size to permit control of the datum requirements.

DATUM FEATURE SYMBOL

The *datum feature symbol* is placed on the drawing to identify the features of the object that are specified as datums and referred to as datum features. This symbol may be placed in the following places on the drawing:
❑ The view where a surface appears as an edge.
❑ The extension line projecting from the edge view of a surface.
❑ On a chain line next to a partial datum surface.
❑ In connection with a diameter or symmetrical dimension when associated with a centerline or center plane.
❑ Attached to a feature control frame.

Datum feature symbols are commonly drawn using thin lines with the symbol size related to the drawing lettering height. The triangular base on the datum feature symbol may be filled or unfilled, depending on the specific departmental preference. The filled base helps easily locate these symbols on the drawing. Example 3-1 shows the specifications for drawing a datum feature symbol.

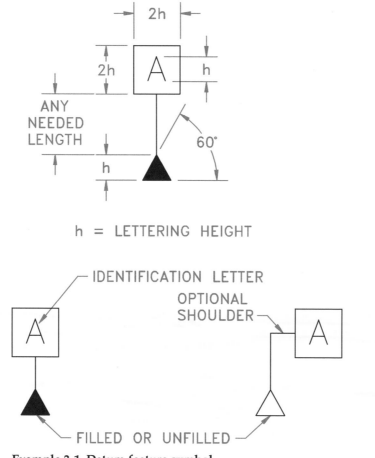

Example 3-1. Datum feature symbol.

DATUM FEATURE SURFACE

The *datum feature* is the *actual* feature of the part that is used to establish the datum. When the datum feature is a surface, it is the actual surface of the object that is identified as the datum. Look at the magnified view of a datum feature placed on the simulated datum in Example 3-2. Observe the following items:

Datum feature: The actual surface of the part.

Simulated datum: The plane established by the inspection equipment such as a surface plate or inspection table.

Datum plane: The theoretically exact plane established by the true geometric counterpart of the datum feature.

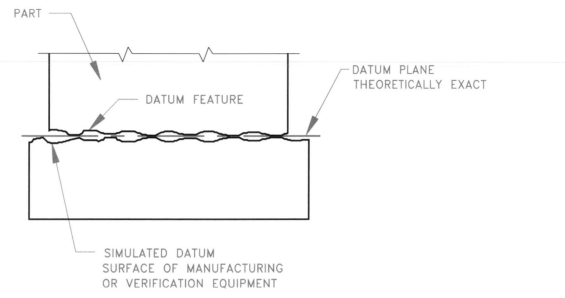

Example 3-2. Datum plane, datum feature, and the simulated datum.

When a surface is used to establish a datum plane on a part, the datum feature symbol is placed on the edge view of the surface or on an extension line in the view where the surface appears as a line. Refer to Example 3-3. A leader line may also be used to connect the datum feature symbol to the view in some applications.

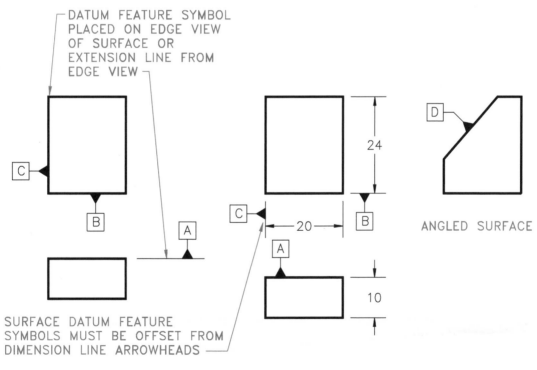

Example 3-3. The datum feature symbol is placed on the edge view or on an extension line in the view where the surface appears as a line.

Geometric Control of Datum Surface

The datum surface may be controlled by a geometric tolerance such as flatness, straightness, circularity, cylindricity, or parallelism. Measurements taken from a datum plane do not take into account any variations of the datum surface from the datum plane. Any geometric tolerance applied to a datum should only be specified if the design requires the control. Example 3-4 shows the feature control frame and datum feature symbol together.

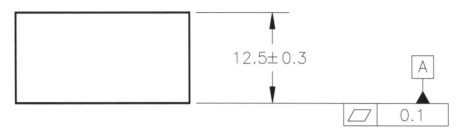

Example 3-4. A feature control frame and datum feature symbol

Example 3-5 is a magnified representation that shows the meaning of the drawing in Example 3-4.

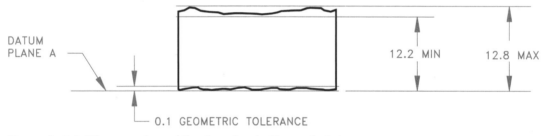

Example 3-5. The meaning of the drawing in Example 3-4.

The geometric tolerance of 0.1 is specified in the feature control frame. The maximum size that the part can be produced is the upper limit of the dimensional tolerance, or MMC. The MMC is 12.5 + 0.3 = 12.8. The minimum size that the part can be produced is the lower limit of the dimensional tolerance, or LMC. The LMC is 12.5 - 0.3 = 12.2.

THE DATUM REFERENCE FRAME CONCEPT

Datum features are selected based on their importance to the design of the part. Generally three datum features are selected that are perpendicular to each other. These three datums are called the *datum reference frame*. The datums that make up the datum reference frame are referred to as the *primary datum, secondary datum,* and *tertiary (third) datum.* As their names imply, the primary datum is the most important, followed by the other two in order of importance. Refer to Example 3-6 and notice how the direction of measurement is projected to various features on the object from the three common perpendicular planes of the datum reference frame.

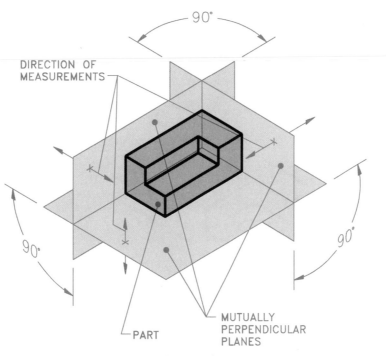

DIRECTION OF
MEASUREMENTS

90°

90°

90°

MUTUALLY
PERPENDICULAR
PLANES

PART

Example 3-6. The datum reference frame.

The datum feature symbols on the drawing relate to the datum features on the part. Notice datum feature symbols A, B, and C as you look at Example 3-7. Also, notice the datum reference order A, B, C in the feature control frame. The datum reference in the feature control frame tells you that Datum A is primary, Datum B is secondary, and Datum C is tertiary. An explanation of the symbols on the drawing in Example 3-7 follows.

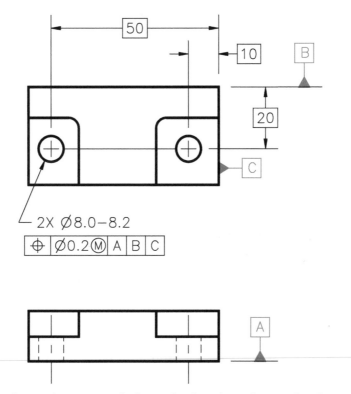

Example 3-7. The datum feature symbols on the drawing relate to the datum features on the part. Look at this drawing as you refer to Examples 3-8 through 3-10 and the related discussion in the text.

The surface of the part labeled as the primary datum is placed on the surface of an inspection table, or manufacturing inspection equipment, as shown in Example 3-8. Now measurements can be made from the primary datum inspection table surface to features that are dimensioned from the primary datum.

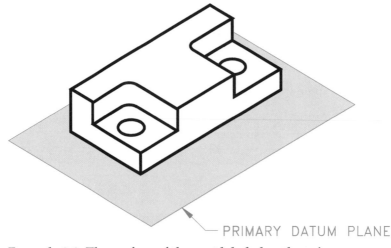

PRIMARY DATUM PLANE

Example 3-8. The surface of the part labeled as the primary datum is placed on the surface of the inspection table.

The part is then positioned against the secondary datum, as shown in Example 3-9. With the part held against the primary and secondary datums, dimensions can be verified from the secondary datum inspection table surface to features that are dimensioned from the secondary datum.

SECONDARY DATUM PLANE

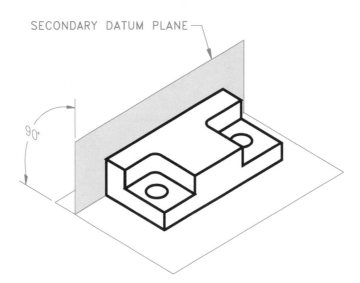

90°

Example 3-9. The part is now positioned against the secondary datum.

Finally, the tertiary datum is established to totally confine the part in the datum reference frame, as shown in Example 3-10. Now with the part totally confined in the datum reference frame, every measurement made from the simulated datum planes to related features on the part are reliable and have the same origin every time.

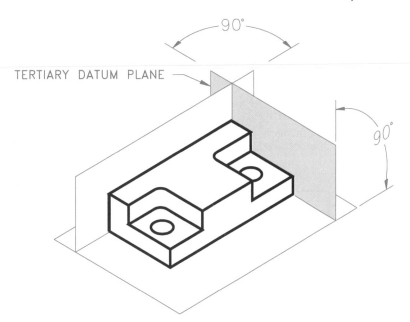

Example 3-10. Finally, the tertiary datum is established to totally confine the part in the datum reference frame.

The surfaces of this inspection equipment are the simulated datums. The datums are the assumed exact planes established by the true geometric counterpart of the datum features. Measurements cannot be made from the datums, because they are only assumed to exist and do not in actuality. The machine tables, surface plates, or inspection tables are of such high quality that they are used to act as the datums where measurements are taken from and where dimensions are verified. In this way each dimension always originates from the same reliable location. Dimensions are never taken or verified from one surface of the part to another. Dimensions always originate from the datum planes.

Refer again to Example 3-7 for review. Notice again the feature control frame associated with the 2X ∅8.0-8.2 dimension. The last three compartments in the feature control frame provide the datum reference. This is known as the datum *order of precedence.* The primary datum (A) is given first followed by the secondary (B) and tertiary (C). For instructional purposes, this example labels datum feature symbols conveniently as A, B, and C. In industry and later in this book, other letters are used to identify datums such as D, E, F, or L, M, N, or X, Y, Z. The letters O, Q, and I are avoided because they may resemble numbers.

As the part is positioned on the datum reference frame, as illustrated in Example 3-8 through Example 3-10, there must be three points anywhere on the primary datum feature in contact with the first datum plane. At least two points of contact are required to establish the secondary datum feature against its datum plane. At least one point must contact the datum plane on the tertiary datum feature. These points of contact, referred to as *high points,* take into account possible irregularity in the manufacture of the part within design limits. Positioning the part in the datum reference frame in this manner ensures a common basis for measurements.

Multiple Datum Reference Frame

Depending on the functional requirements of a part, more than one datum reference frame may be established. In Example 3-11, datums X, Y, and Z constitute one datum reference frame, while datums L and M establish a second reference frame. The relationship between the two datum reference frames is controlled by the angularity tolerance on datum feature L. Datum M is the axis of the large hole that the datum feature symbol is connected to. Datum axes are discussed later in this chapter.

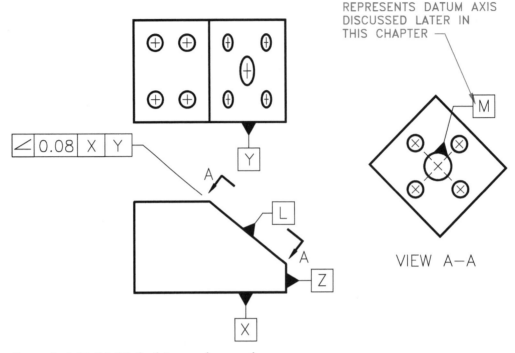

Example 3-11. Multiple datum reference frame.

DATUM TARGET SYMBOLS

In many situations it is not possible to establish an entire surface, or entire surfaces, as datums. When this happens, then datum targets may be used to establish datum planes. This procedure is especially useful on parts with surface or contour irregularities, such as some sheet metal, sand cast, or forged parts that are subject to bowing or warpage. This method can also be applied to weldments where heat may cause warpage. *Datum targets* are designated points, lines, or surface areas that are used to establish the datum reference frame. The datum target symbol is drawn as a circle using thin lines. The circle is divided into two parts with a horizontal line. The bottom half provides the datum reference letter and the specific datum target number on that datum. The top half is left blank if a datum target point or line is identified. When identifying a datum target area, the top half contains the diameter of the area. The dimension for the datum target area may be placed outside the datum target symbol with a leader and a dot pointing to the upper half (if the diameter symbol of the dimension is to big to fit inside). See Example 3-12. The datum target symbol is connected with a leader to the datum target point, line, or area, as shown in Example 3-13.

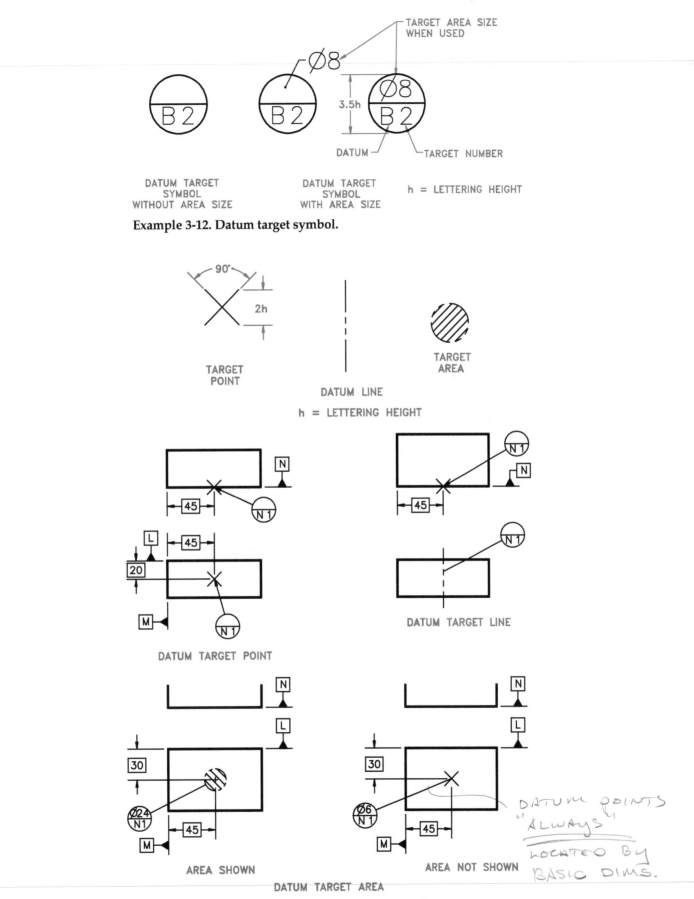

Example 3-12. Datum target symbol.

Example 3-13. The datum target point, datum target line, datum target area, and examples of use.

Datum Target Points

Datum planes are established by the datum points as follows:

The *primary datum plane* must be established by at least three points on the primary datum surface. These points are used to provide stability on the primary plane, similar to a three legged stool.

The *secondary datum plane* must be located by at least two points on the related secondary datum surface. Two points provide the required stability for the secondary plane.

The *tertiary datum plane* must be located by at least one point on the related tertiary datum surface. One point of contact at the tertiary datum plane is all that is required to complete the datum reference frame and provides complete stability of the part in the datum reference frame.

Datum or chain dimensioning may be used to locate datum target points. The location dimensions must originate from datums. Datum target points are established on the drawing using basic or tolerance dimensions. Established tooling or gaging tolerances apply when datum targets are located with basic dimensions. Datum targets are established on the part with fixtures and with pins. These pins contact the part where the datum targets are specified. Example 3-14 shows a pictorial drawing of the datum target points on the primary, secondary, and tertiary datums.

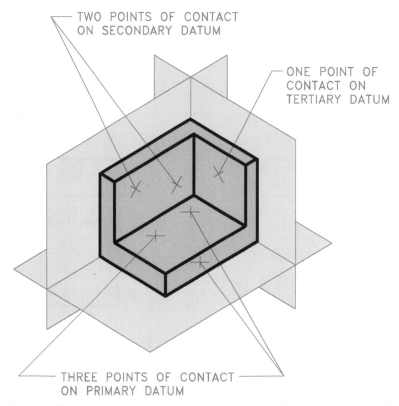

Example 3-14. Datum target points on primary, secondary, and tertiary datums.

As you have seen, each datum target point is identified with a datum target symbol. The information inside the datum target symbol identifies the datum target point, as shown in Example 3-15.

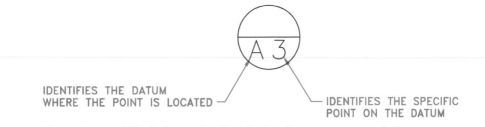

Example 3-15. The information inside the datum target symbol identifies the datum target point, target line, or target area.

The datum points may be located with basic dimensions or tolerance dimensions. Refer to Example 3-16 for a multiview representation using basic dimensions to locate the datum target points. The datum feature symbols appear on the view where the datum surface is a line and the datum points are located on the surface view of the related datum. The datum target symbol may be placed on the view where the surface appears as an edge if the drawing arrangement dictates such placement, as shown in Example 3-17.

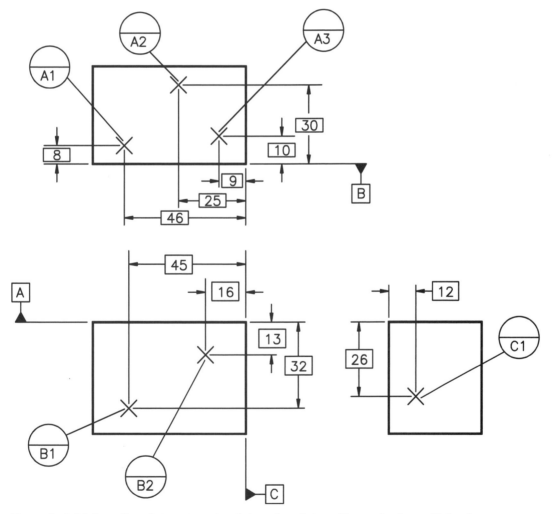

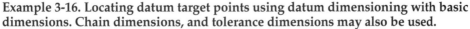

Example 3-16. Locating datum target points using datum dimensioning with basic dimensions. Chain dimensions, and tolerance dimensions may also be used.

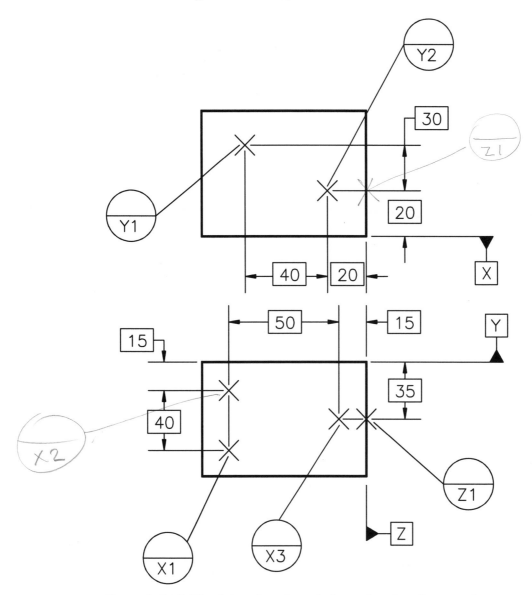

Example 3-17. The datum target symbol may be placed on an edge when a surface view is not available, such as point Z1 in this drawing. Note: Datum target point Z1 is located at the corner where datum surfaces Y and Z meet. Otherwise, there would be a location dimension in the top view.

When datum target points are used on a drawing to identify a datum plane, the datum plane is established by locating pins at the datum tangent points as shown in the magnified representation in Example 3-18. The locating pins are rounded or pointed standard tooling hardware.

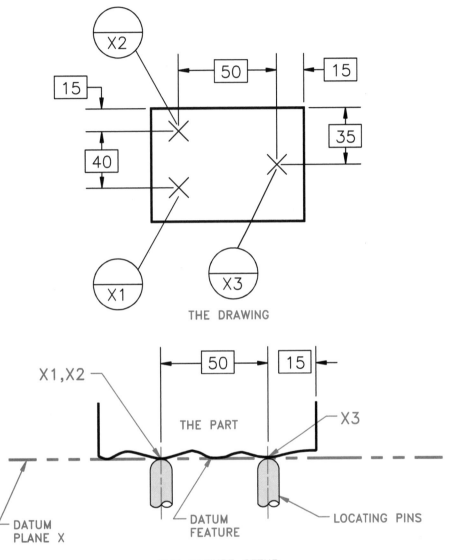

THE DRAWING

THE FIXTURE SETUP

Example 3-18. Datum target points on a drawing and the points established with locating pins.

Datum Target Areas

Areas of contact may also be used to establish datums. When this is done, the shape of the datum target area is outlined by phantom lines with section lines through the area. Circular areas are dimensioned with basic or tolerance dimensions to locate the center. The diameter of the target area is provided in the upper half of the datum target symbol or with a leader and dot pointing to the upper half, as shown in Example 3-19. The locating pins for target areas are flat end tooling pins with the pin diameter equal to the specified size of the target area.

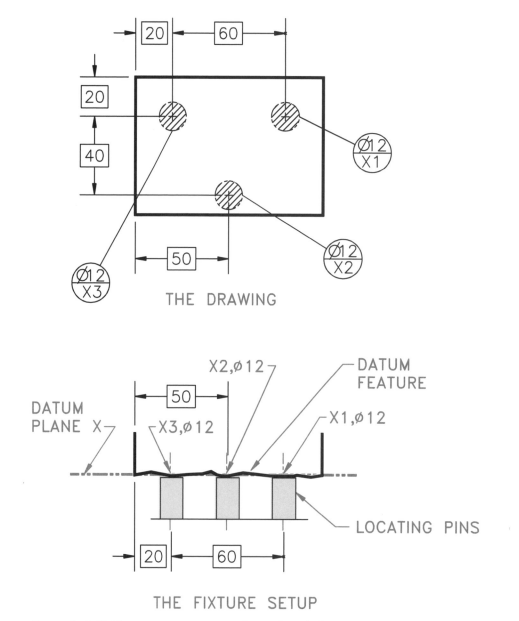

Example 3-19. Datum target areas are located to their centers. The locating pins for target areas are flat end tooling pins with the pin diameter equal to the specified size of the target area.

When the area is too small to accurately or clearly display on a drawing, then a datum target point is used at the center location. The top half of the datum target symbol identifies the diameter of the target area, as shown in Example 3-20.

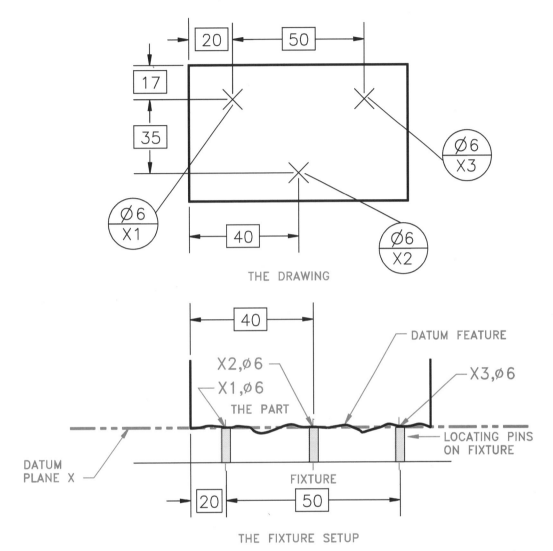

THE DRAWING

THE FIXTURE SETUP

Example 3-20. When the datum target area is too small to show, the datum target point is used and the target area size is given in the top half of the datum target symbol.

Datum Target Lines

A *datum target line* is indicated by the target point symbol "X" on the edge view of the surface and by a phantom line on the surface view. Refer to Example 3-21. If the locating pins are cylindrical, then the datum target line is along the tangency where the pins meet the part. The pins may also be knife-edged. A surface is often placed at 90° to the pin to create the datum reference frame.

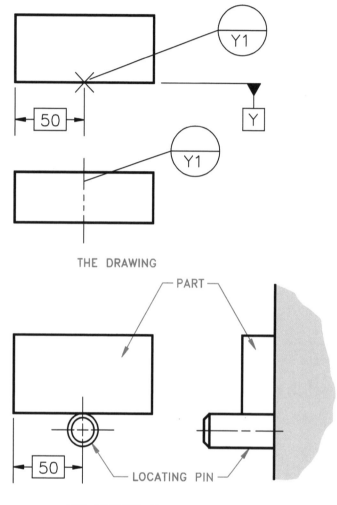

Example 3-21. Datum target line.

PARTIAL DATUM SURFACE

A portion of a surface may be used as a datum. For example, this may be done when a part has a hole or group of holes at one end where it may not be necessary to establish the entire surface as a datum to effectively locate the features. This may be accomplished on a drawing using a chain line dimensioned with basic dimensions to show the location and extent of the partial datum surface. The datum feature symbol is attached to the chain line. The datum plane is then established at the location of the chain line, as shown in Example 3-22.

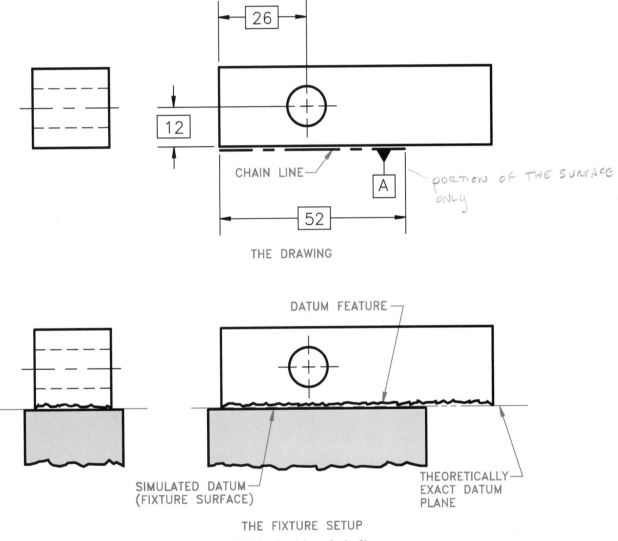

Example 3-22. Partial datum surface established with a chain line.

COPLANAR SURFACE DATUMS

Coplanar surfaces are two or more surfaces that are on the same plane. The relationship of coplanar datum features establishes the surfaces as one datum plane in correlated feature control frame specifications. A phantom line is placed between the surfaces if a void, such as a slot, exists. The phantom line between surfaces is omitted when the area between the surfaces is higher than the datum features. The surfaces are treated as a single, noncontinuous surface. The number of surfaces may be specified in a note, such as "2 SURFACES," below the related feature control frame. The datum reference in the feature control frame gives both datum letters separated by a dash. See Example 3-23. This concept is also discussed in Chapter 5 with an application for profile tolerances of coplanar surfaces.

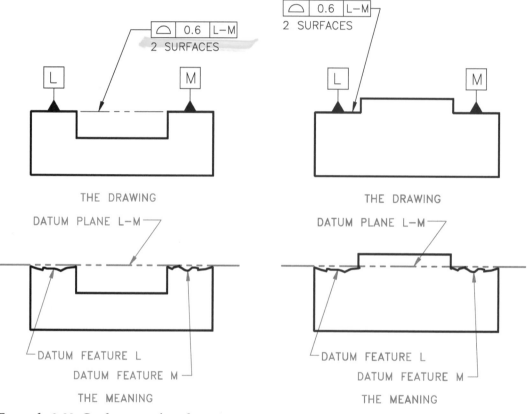

Example 3-23. Coplanar surface datums.

DATUM AXIS

A cylindrical object may be a datum feature. When the cylindrical datum feature is used, the center axis is known as the datum axis. There are two theoretical planes intersecting at 90°. These planes are represented by the centerlines of the drawing. Where these planes intersect is referred to as the *datum axis.* The datum axis is the origin for related dimensions, while the X and Y planes indicate the direction of measurement. A datum plane is added to the end of the object to establish the datum frame, as shown in Example 3-24.

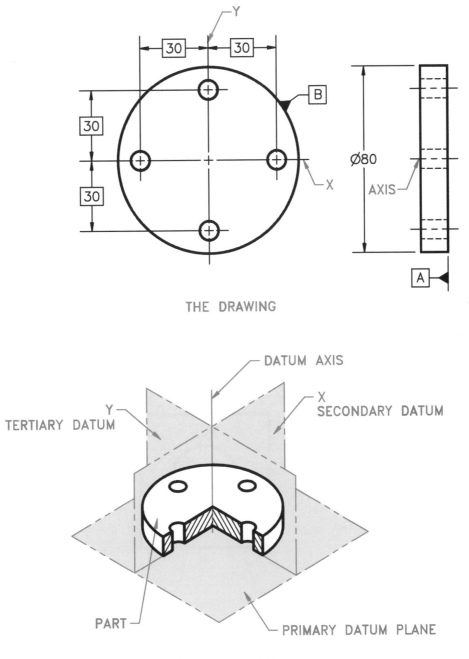

Example 3-24. Datum axis established by datum coordinates.

Placement of the Datum Feature Symbol for a Datum Axis

When the datum is an axis, the datum feature symbol may be placed on the drawing using one of the following methods (also shown in Example 3-25):

❏ The symbol can be placed on the outside surface of a cylindrical feature.
❏ The symbol can be centered on the opposite side of the dimension line arrowhead.
❏ The symbol can replace the dimension line and arrowhead when the dimension line is placed outside of the extension lines.
❏ The symbol can be placed on a leader line shoulder.
❏ The symbol can be placed below, and attached to, the center of a feature control frame.

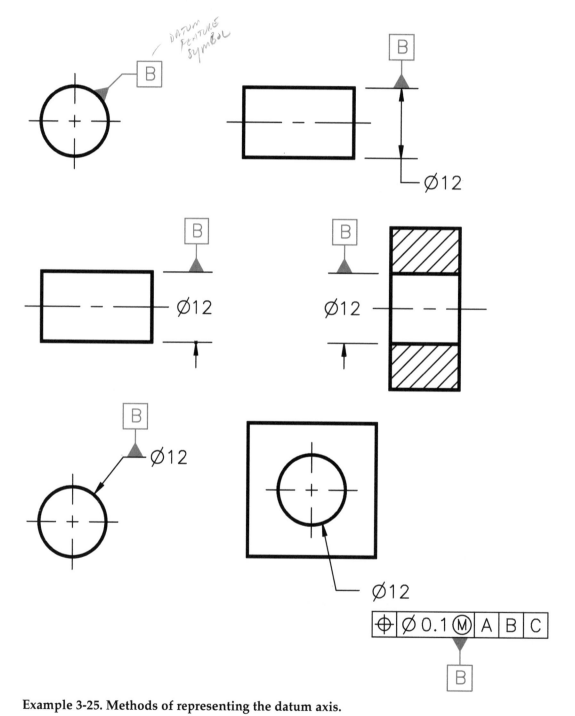

Example 3-25. Methods of representing the datum axis.

Simulated Datum Axis

The *simulated datum axis* is the axis of a perfect cylindrical inspection device that contacts the datum feature surface. For an external datum feature, as shown in Example 3-26, the inspection device is the smallest (MMC) *circumscribed* cylinder. The inspection device for an internal datum feature is the largest (MMC) *inscribed* cylinder, as shown in Example 3-27.

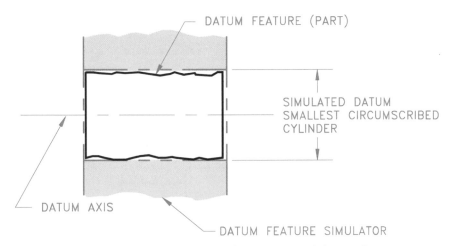

Example 3-26. Simulated datum axis for an external datum feature.

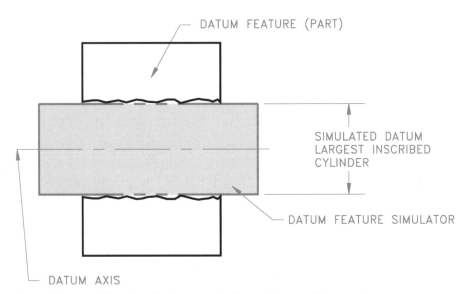

Example 3-27. Simulated datum axis for an internal datum feature.

Coaxial Datum Features

Coaxial means two or more cylindrical shapes that share a common axis. *Coaxial datum features* exist when a single datum axis is established by two datum features that are coaxial. When more than one datum feature is used to establish a single datum, the datum reference letters are separated by a dash and placed in one compartment of the feature control frame. These datum reference letters are of equal importance and may be placed in any order. See Example 3-28. A datum axis established by coaxial datum features is normally used as a primary datum.

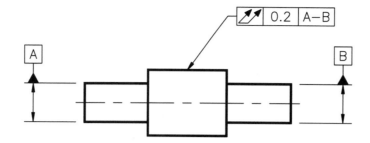

THE DRAWING

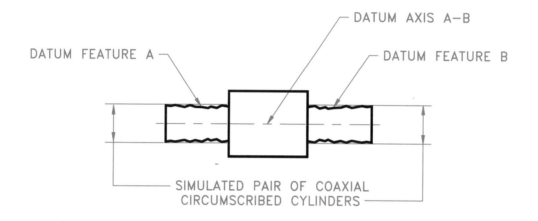

THE MEANING

Example 3-28. Coaxial datum features.

The Datum Axis of Screw Threads, Gears, and Splines

When a screw thread is used as a datum axis, the datum axis is established from the pitch cylinder unless otherwise specified. If another feature of the screw thread is desired, then note "MAJOR DIA" or "MINOR DIA" is placed next to the datum feature symbol.

A specific feature such as the major diameter should be identified when a gear or spline is used as a datum axis. When this is done, the note "MAJOR DIA," "MINOR DIA," or "PITCH DIA" is placed next to the datum feature symbol as appropriate. The use of a screw thread, gear, or spline should be avoided for use as a datum axis unless necessary.

Datum Axis Established with Datum Target Symbols

Datum target points, lines, or surface areas may also be used to establish a datum axis. A primary datum axis may be established by two sets of three equally spaced targets–a set near one end of the cylinder and the other set near the other end, as shown in Example 3-29.

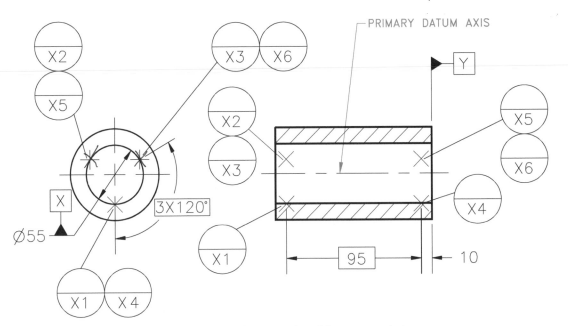

Example 3-29. Establishing a primary datum axis with target points.

When two cylindrical features of different diameters are used to establish a datum axis, then the datum target points are identified in correlation to the adjacent cylindrical datum feature. Refer to Example 3-30.

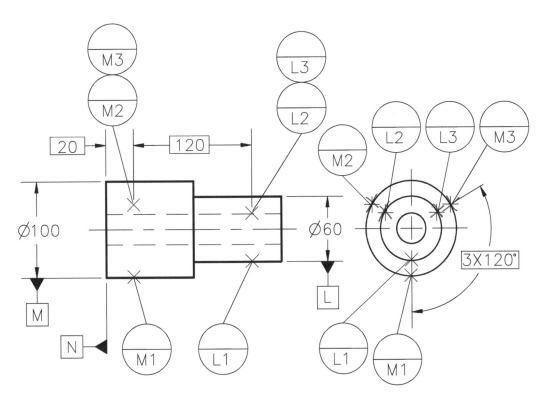

Example 3-30. Two cylindrical features of different diameters used to establish a datum axis.

Cylindrical datum target areas and circular datum target lines may also be used to establish the datum axis of cylindrical shaped parts, as shown in Example 3-31. In this application the datum target area is a designated width band that goes all around the part. This datum target area is shown as two phantom lines with section lines between. The datum target line is a phantom line that goes all around the part.

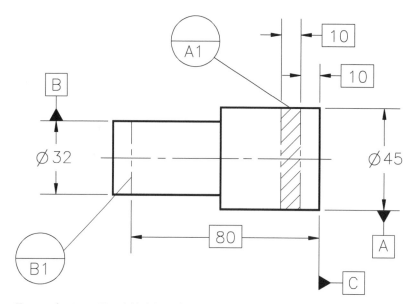

Example 3-31. Establishing datum axes with a cylindrical datum target area and a circular datum target line.

A secondary datum axis may be established by placing three equally spaced targets on the cylindrical surface. Refer to Example 3-32.

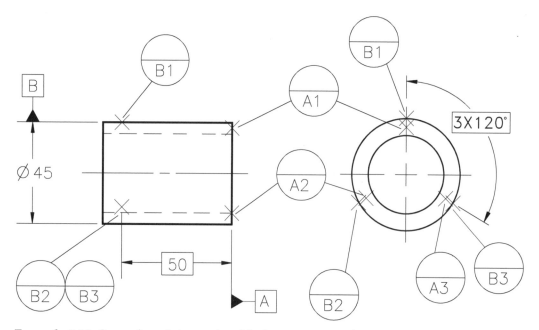

Example 3-32. Secondary datum axis with datum target points.

DATUM CENTER PLANE

Elements on a rectangular shaped symmetrical part or feature may be located and dimensioned in relationship to a datum center plane. The representation and related meaning of datum center plane symbols are shown in Example 3-33.

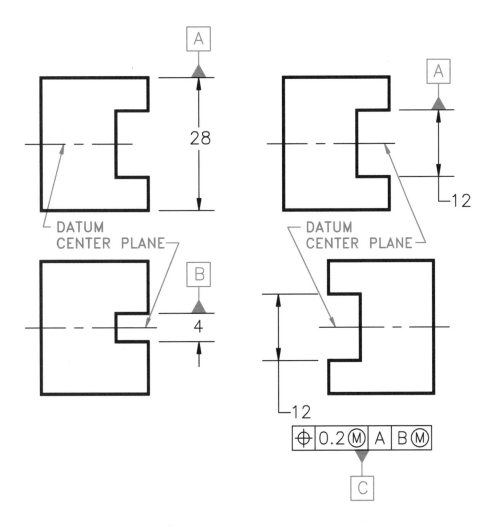

AXIS AND CENTER PLANE DATUM FEATURE
SYMBOLS MUST ALIGN WITH OR REPLACE THE
DIMENSION LINE ARROWHEAD OR BE PLACED ON
THE FEATURE, LEADER SHOULDER, OR FEATURE
CONTROL FRAME.

Example 3-33. Placement of center plane datum feature symbols.

Be sure you notice the difference between the datum feature symbol associated with the datum axis discussed earlier and the datum center plane. The drawings look similar. The datum axis is on a cylindrical feature such as an external shaft or an internal hole. Look carefully at the drawing views and notice that the cylindrical datum feature is dimensioned with a diameter (refer back to Example 3-25). The datum center plane is the plane that splits a symmetrical feature such as a slot or tab. The dimension associated with the center plane datums do not have a diameter symbol because they are not round. The datum feature symbol is placed one of the following ways:

❏ Centered on the opposite side of the dimension line arrowhead.
❏ Replaces the dimension line and arrowhead when the dimension line is placed outside of the extension lines.
❏ Placed on a leader shoulder.
❏ Placed below and attached to the center of a feature control frame.

Simulated Datum Center Plane

The simulated datum center plane is the center plane of a perfect rectangular inspection device that contacts the datum feature surface. For an external datum feature the datum center plane is established by two parallel planes at minimum (MMC) separation, as shown in Example 3-34. For an internal datum feature, the datum center plane is established by two parallel planes at maximum (MMC) separation, as shown in Example 3-35.

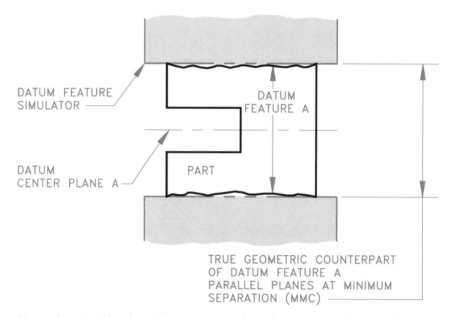

Example 3-34. Simulated datum center plane for an external datum feature.

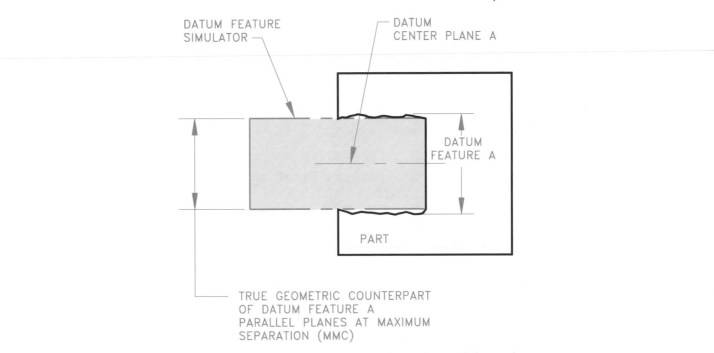

Example 3-35. Simulated datum center plane for an internal datum feature.

THE CENTER OF A PATTERN OF FEATURES AS THE DATUM AXIS

The center of a pattern of features, such as the holes in the part shown in Example 3-36, may be specified as the datum axis when the datum feature symbol is placed under, and attached to, the middle of the feature control frame. In this application, the datum axis is the center of the holes as a group. This will be discussed further in Chapter 7 in regard to location tolerances.

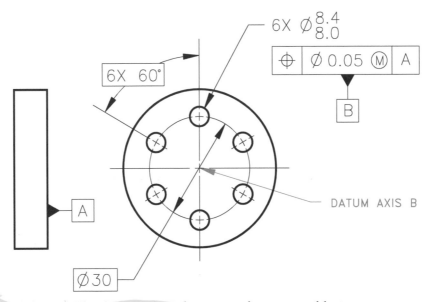

Example 3-36. The datum axis at the center of a pattern of features as a group.

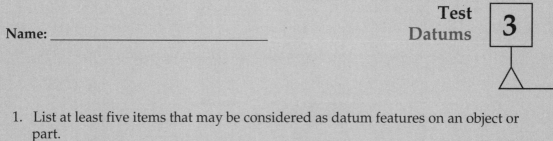

Name: _____

Test
Datums   3

1. List at least five items that may be considered as datum features on an object or part.

   a) _VIEW WHERE SURFACE APPEARS AS AN EDGE_

   b) _EXTENTION LINE_

   c) _ON CHAIN LINE @ PARTIAL DATUM PLANE_

   d) _CYLINDRICAL AXIS_

   e) _ATTACHED TO A DATUM FEATURE FRAME_

2. A _DATUM_ plane is a theoretically exact plane.

3. A datum surface or feature is the _ACTUAL_ surface of an object that is used to establish a _DATUM_ plane.

4. Identify the datum feature, the part, the simulated datum, and the datum plane on the following illustration.

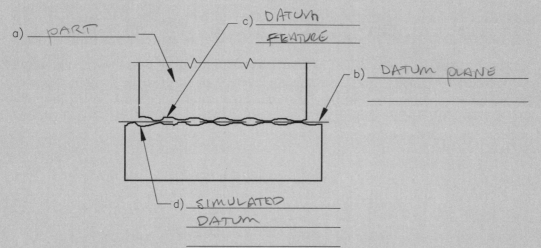

a) _PART_

c) _DATUM FEATURE_

b) _DATUM PLANE_

d) _SIMULATED DATUM_

5. Complete this statement: When a datum surface is used on a part, the datum feature symbol is placed _ON THE EDGE VIEW OF THE PART OR ON AN AN EXTENSION LINE IN THE VIEW WHERE_. _THE SURFACE APPEARS AS A LINE_

6. Given the following drawing and related meaning, fill in the blanks at Part I a and b and at Part II a, b, c, and d. Provide the actual dimensions as related to *the drawing* at Part II b, c, and d.

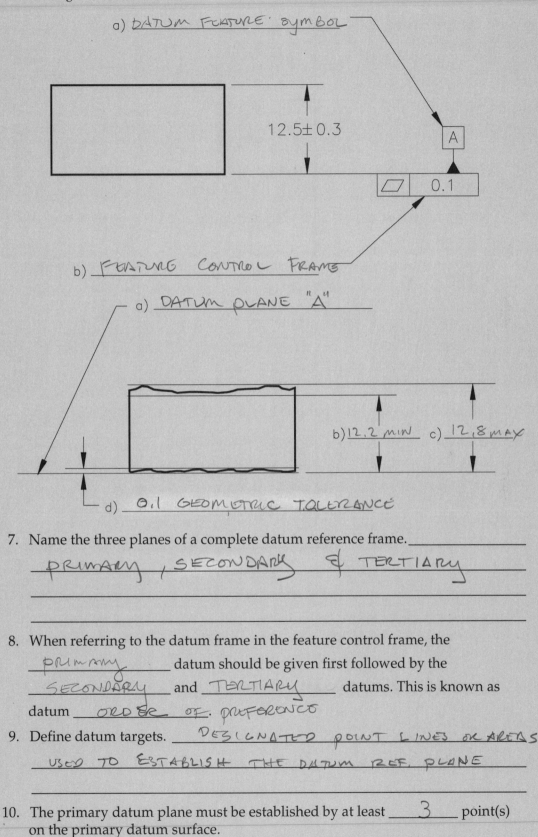

a) DATUM FEATURE SYMBOL

12.5±0.3

A

0.1

b) FEATURE CONTROL FRAME

a) DATUM PLANE "A"

b) 12.2 MIN   c) 12.8 MAX

d) 0.1 GEOMETRIC TOLERANCE

7. Name the three planes of a complete datum reference frame. _____
    PRIMARY , SECONDARY & TERTIARY
    _____
    _____

8. When referring to the datum frame in the feature control frame, the
    ___PRIMARY___ datum should be given first followed by the
    ___SECONDARY___ and ___TERTIARY___ datums. This is known as
    datum ___ORDER OF PREFERENCE___

9. Define datum targets. ___DESIGNATED POINT LINES OR AREAS___
    ___USED TO ESTABLISH THE DATUM REF. PLANE___
    _____

10. The primary datum plane must be established by at least ___3___ point(s) on the primary datum surface.

11. The secondary datum plane must be established by at least ___2___ point(s) on the related secondary datum surface.

12. The tertiary datum plane must be established by at least ___1___ point(s) on the related tertiary datum surface.

13. How are datum target areas represented on a drawing? _____
    ___ PHANTOM CIRCLE W/CROSS HATCHING ___
    _____
    _____
    _____

14. The circular datum target area is dimensioned with ___BASIC___ dimensions to locate the ___CENTER___ from datums and the diameter of the area is provided in the ___UPPER___ half of the datum target symbol.

15. How are datum target areas treated on a drawing when the target area is too small to draw? ___ AN "X" @ THE DATUM TARGET LOCATION ___
    _____
    ___ (UPPER HALF OF SYMBOL SHOWS SIZE OF AREA) ___
    _____
    _____
    _____

16. How are datum target lines represented on a drawing? ___ "X" @ VIEW ___
    ___ WHERE LINE APPEARS AS A POINT & PHANTOM ___
    ___ LINE IN SURFACE VIEW ___
    _____
    _____
    _____
    _____

17. When a portion of a surface is used as a datum, this is referred to as a(n) ___PARTIAL___ datum surface.

18. When more than one datum feature is used to establish a single datum, this is referred to as ___CO PLANAR___ datum features.

19. Depending on the functional requirements of a part, more than one datum reference frame may be established. This is referred to as a(n) ___MULTIPLE___ datum reference frame.

20. Define coplanar surfaces. *SURFACES THAT LIE IN THE SAME PLANE*

21. How is a datum axis represented on a drawing? ① *ON THE OUTSIDE SURFACE OF A CYL. FEATURE* ② *OPP. SIDE OF DIM. LINE ARROWHEAD* ③ *REPLACE DIM. LINE WHEN OUTSIDE EXTENSION LINE* ④ *PLACED ON A LEADER LINE SHOULDER* ⑤ *PLACED BELOW & ATTACHED C CTR. OF FEATURE CNTL FRAME*

22. Label the elements a, b, c, and d below that represent the fixture setup for a datum axis.

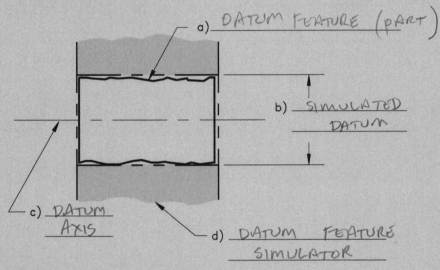

a) *DATUM FEATURE (PART)*

b) *SIMULATED DATUM*

c) *DATUM AXIS*

d) *DATUM FEATURE SIMULATOR*

23. A primary datum axis may be established by two sets of three equally spaced target points. True or False? *TRUE*

24. Cylindrical datum target areas and circular datum target lines may be used to establish the datum axis of cylindrical shaped parts. True or False? *TRUE*

25. How is a datum center plane shown on a drawing? _____ *p 86*
① *CENTERED ON OPP. SIDE OF DIM. LINE ARROWHEAD*
② *REPLACE DIM. LINE & ARROWHD IF OUTSIDE EXTENS. LINE*
③ *PLACED ON A LEADER SHOULDER*
④ *PLACED BELOW & ATTACHED TO CTR. OF FEATURE CONTROL FRM.*

26. Identify the items in the drawings below labeled a through k.

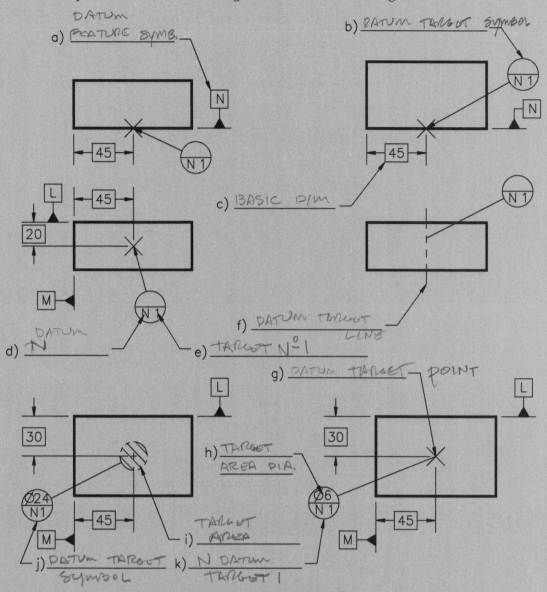

DATUM
a) FEATURE SYMB.

b) DATUM TARGET SYMBOL

c) BASIC DIM

DATUM
d) N

e) TARGET Nº 1

f) DATUM TARGET LINE

g) DATUM TARGET POINT

h) TARGET AREA DIA.

i) TARGET AREA

j) DATUM TARGET SYMBOL

k) N DATUM TARGET 1

27. On the blank lines (a through k) below each of the following drawings identify if the datum feature symbols represent datum surface, datum axis, or datum center plane.

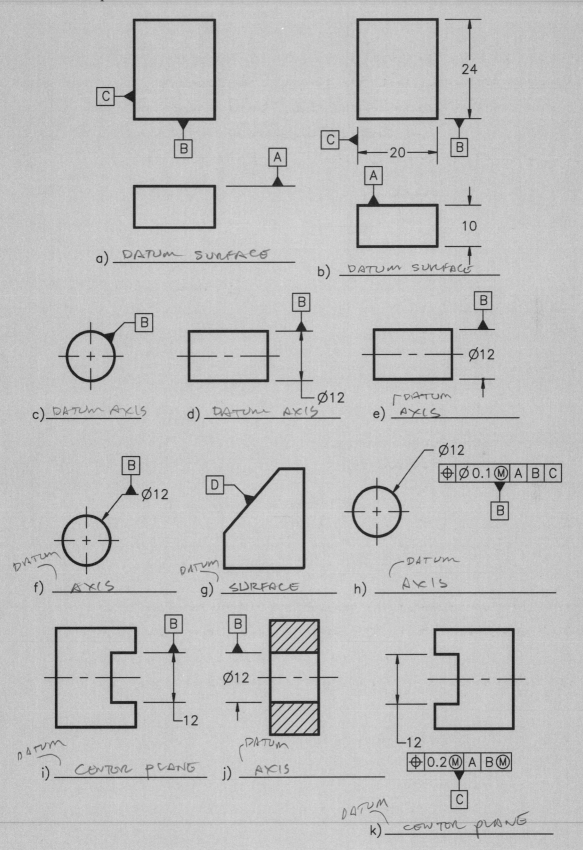

a) __DATUM SURFACE__

b) __DATUM SURFACE__

c) __DATUM AXIS__

d) __DATUM AXIS__

e) __DATUM AXIS__

f) __DATUM AXIS__

g) __DATUM SURFACE__

h) __DATUM AXIS__

i) __DATUM CONTOR PLANE__

j) __DATUM AXIS__

k) __DATUM CONTOR PLANE__

28. Label the elements a, b, c, d, and e below that represent the fixture setup for a datum center plane.

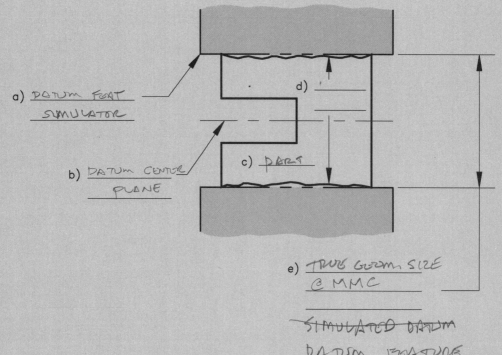

a) <u>DATUM FEAT</u>
   <u>SIMULATOR</u>

b) <u>DATUM CENTER</u>
   <u>PLANE</u>

c) <u>PART</u>

d)

e) <u>TRUE GEOM SIZE</u>
   <u>@ MMC</u>

~~SIMULATED DATUM~~
DATUM FEATURE
SIMULATOR

Print Reading Exercise

Name: _____

The following print reading exercise uses actual industry prints with related questions that require you to read specific dimensioning and geometric tolerancing representations. The answers should be based on previously learned content of this book. The prints used are based on ASME standards, however company standards may differ slightly. When reading these prints, or any other industry prints, a degree of flexibility may be required to determine how individual applications correlate with the ASME standards.

Refer to the print of the SLEEVE-DEWAR REIMAGING found on page 303.

1. Describe Datum A. *LEFT END OF SLEEVE*

2. Describe Datum B. *CENTER AXIS OF SLEEVE*

3. Give the location dimension of the Ø.107±.001 hole from Datum A. *.469"*

Refer to the print of the BRACKET found on page 304.

4. Identify the datums that make up the primary datum reference frame. _____
   A, B, C

5. Describe Datum D. *AXIS OF .875" Ø HOLE*

6. Identify the primary, secondary, and tertiary datum references associated with the positional geometric tolerance placed with the Ø.875±.005 dimension:
   Primary *A   BACK FACE*
   Secondary *B   LH EDGE*
   Tertiary *C   BOT. EDGE*

7. Give the location dimensions to the three Ø.437±.005 holes from Datum C.
   .900, 1,800, 2,700

8. Give the location dimensions to the three Ø.437±.005 holes from Datum B.
   4.050

9. Give the location dimensions to the four Ø.187±.003 holes from Datum A. (Note that 2 of the holes are dimensioned from the surface opposite Datum A.)
   .374, .374 – .093

10. Give the location dimensions to the four Ø.187±.003 holes from Datum C.
    .600, 3,000

11.   Are the location dimensions in question number 7 placed using datum or chain
      dimensioning? _____DATUM_____CHAIN_____

Refer to the print of the HUB-STATIONARY ATU found on page 305.

12.   Describe Datum A. _____
      _R.H. FACE OF RH RIB OF 4.812" ⌀_____
      _____
      _____
      _____

13.   Describe Datum B. _____
      ____LONGITUDINAL AXIS OF PART_____
      _____
      _____
      _____

14.   Describe Datum C. _____
      ____LONGITUDINAL AXIS OF PART_____
      _____
      _____
      _____

15.   Give the dimension from Datum A to the left face of the part. _6.673/6.678_

16.   Give the horizontal and vertical location dimensions from Datum C to the six
      ⌀.352+.005/-.001 holes. __.763" x 1.321 & 1.525 x 0__
      _____
      _____

Refer to the print of the PEDAL-ACCELERATOR found on page 306.

17.   How many points of contact are used to establish the following datums?
      Datum A __3__, Datum B __2__, Datum C __1__.

18.   Identify the following items for each of the datum target areas: the datum
      reference, specified number on the datum, and the area size and shape. _____
      _____
      _____
      _____

19.   What is the distance between datum target point B-1 and B-2? _105 mm BASIC_

20.   What does the box around the 74.00 dimension mean?_____
      ____BASIC DIM)_____

21. Describe Datum D. _____

    ____AXIS OF 19.10" ⌀ HOLE_____

    _____

    _____

    _____

22. Give the location dimensions between the datum target areas A1 and A2.

    ____112.9 HORIZ X 74.00 VERT X 40.00 Z DIRECT.____

23. Give the location dimensions between the datum target area A1 and datum target point A3. ____165 H, X 58 V. X 40.00 Z____

24. Give the location dimensions between the datum target points A3 and C1.

    ____58 V. X 177.76 X 30.00 Z____

25. Give the location dimensions between the datum target points B1, B2, and C1.

    B1 12.5 V. X 147.76 X 37.00

    B2 12.5 V. X 42.76 X 37.00

    _____

Refer to the print of the DOUBLE V-BLOCK found on page 310.

26. Identify the primary datum reference frame. ____A, B, C____

27. Describe Datum D. ____DATUM CENTER PLANE____

    ____LOCATED @ 30.16" FROM DATUM A.____

    _____

    _____

    _____

28. Give the dimension from Datum C to the bottom of the 1.60 wide slot.

    ____42.88____

29. Give the location dimension to the Ø12.70/12.65 feature from Datum B.

    ____69.85____

30. Give the location dimension to the Ø12.70/12.65 feature from Datum C.

    ____12.70____

Refer to the print of the BRACKET ASSY-EL GIMBAL found on page 315.

31. Give the location dimension from Datum D to the three Ø.109+.004/-.001 COUNTERSINK Ø.172×100° features. ____.125____

32. Describe Datum B. ____AXIS OF 2.49" RAD____

    _____

    _____

    _____

    _____

33. Describe Datum C. _____ AXIS OF .375" RAD _____

_____

_____

_____

_____

34. What is the wall thickness of the material? _____ .375 - .310 = .065 _____

35. Give the vertical location dimension from Datum B to the three .086-56UNC-2B
     features. _____ 2.615 × 2.49 _____

GENL NOTES
⌀.75 × .028 WALL

Chapter 4
Material Condition Symbols

Material Condition Symbols are used in conjunction with the geometric tolerance or datum reference in the feature control frame. The *material condition symbols* establish the relationship between the size of the feature within its given dimensional tolerance and the geometric tolerance. The use of different material condition symbols alters the effect of this relationship. Material condition symbols have been referred to as modifiers, because they modify the geometric tolerance in relationship to the actual produced size of the feature. The *actual produced size* is the measured size after production. The material condition modifying elements are:

❏ Maximum Material Condition, abbreviated MMC.
❏ Regardless of Feature Size, abbreviated RFS.
❏ Least Material Condition, abbreviated LMC.

The material conditions symbols are detailed in Example 4-1.

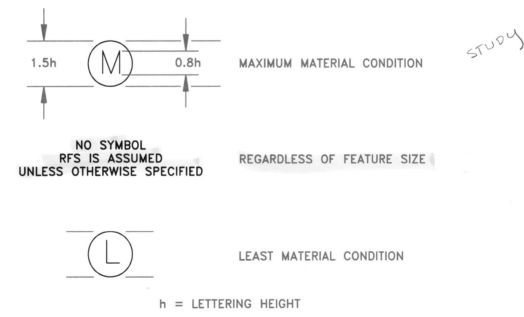

STUDY

Example 4-1. Material condition symbols.

Notice in Example 4-1 that there is no symbol for Regardless of Feature Size (RFS). RFS is assumed for all geometric tolerance applications and related datum references unless otherwise specified. The MMC or LMC symbols must be in the feature control frame if these applications are intended. The material condition symbols, when used, are placed after the geometric tolerance or datum reference, as shown in Example 4-2.

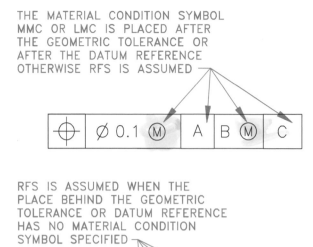

THE MATERIAL CONDITION SYMBOL
MMC OR LMC IS PLACED AFTER
THE GEOMETRIC TOLERANCE OR
AFTER THE DATUM REFERENCE
OTHERWISE RFS IS ASSUMED

RFS IS ASSUMED WHEN THE
PLACE BEHIND THE GEOMETRIC
TOLERANCE OR DATUM REFERENCE
HAS NO MATERIAL CONDITION
SYMBOL SPECIFIED

Example 4-2. The material condition symbols MMC or LMC, when used, are placed after the geometric tolerance and datum reference in the feature control frame. If no material condition symbol is used, RFS is assumed. This is referred to as "Rule 2" in ASME Y14.5M.

CONVENTIONAL TOLERANCE

Use of the term *conventional tolerancing* in this text refers to tolerances related to dimensioning practices without regard to geometric tolerancing. The limits of a size dimension determine the given variation allowed in the size of the feature. The part shown in Example 4-3 has a Maximum Material Condition of 6.5 and a Least Material Condition of 5.5. The MMC and LMC produced sizes represent the limits of the dimension. The actual part may be manufactured at any size between the limits. Some possible produced sizes in 0.1 mm increments are shown in the chart in Example 4-3.

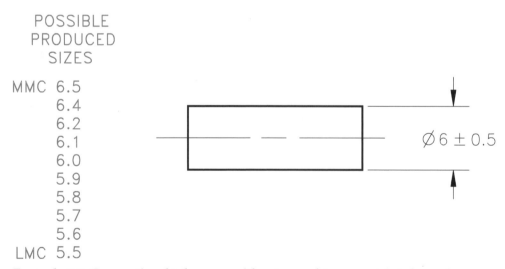

POSSIBLE
PRODUCED
SIZES

MMC 6.5
    6.4
    6.2
    6.1
    6.0
    5.9
    5.8
    5.7
    5.6
LMC 5.5

$\varnothing 6 \pm 0.5$

Example 4-3. Conventional tolerance, without regard to geometric tolerancing.

LIMITS OF SIZE

Conventional tolerancing, without the addition of geometric tolerancing, permits a degree of variation in form, profile, or location because of the tolerance. The degree of form or location control can be increased or decreased by altering the tolerance. The amount of form control implied in a conventional tolerance is determined by the actual size of the feature or part that must be within the given tolerance at any cross section.

The *limits of size* of a feature control the amount of variation in size and geometric form. This is referred to as "Rule 1" in ASME Y14.5M. The limits of size is the boundary between maximum material condition (MMC) and least material condition (LMC). The form of the feature may vary between the upper limit and lower limit of a size dimension. This is known as the *extreme form variation.*

The limits of size and representative extreme form variation of the part shown in Example 4-3 and a mating collar is shown in Example 4-4.

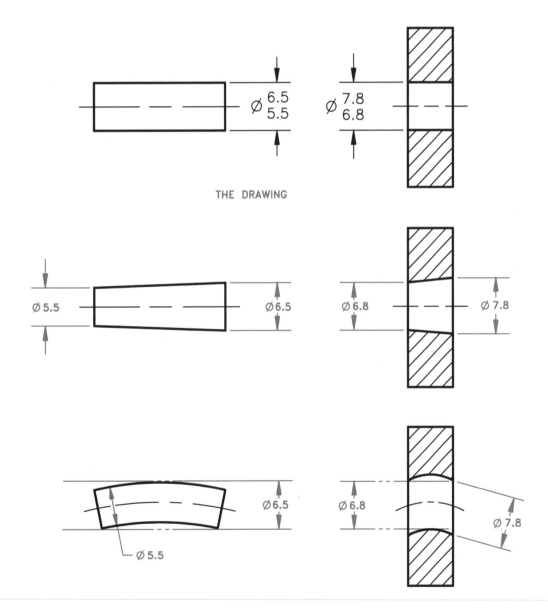

THE DRAWING

POSSIBLE EXTREME FORM VARIATIONS

Example 4-4. The limits of size and representative extreme form variation of the part. This is referred to as "Rule 1" in ASME Y14.5M.

The form of an individual feature is controlled by its size limits in the following ways:

❑   The surface of a feature may not extend beyond the MMC boundary. See the PERFECT FORM BOUNDARY section in this chapter.

❑   When the actual size of a feature departs from MMC, a variation in form is allowed equal to the difference from MMC.

❑   If a feature is produced at LMC, the geometric form may vary between the LMC and MMC boundaries.

PERFECT FORM BOUNDARY

The form of a feature is controlled by the size tolerance limits, as shown in Example 4-5. The boundary of these limits are established at MMC. The *perfect form boundary* is the true geometric form of the feature at MMC. Therefore, if the part is produced at MMC, it must be at perfect form. If a feature is produced at LMC, the form tolerance is allowed to vary within the geometric tolerance zone to the extent of the MMC boundary.

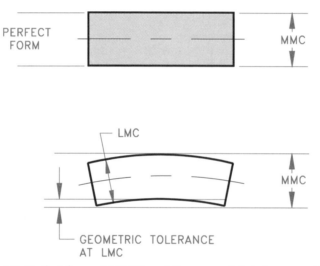

Example 4-5. Perfect form at MMC, and the geometric tolerance at LMC.

In some applications it may be desirable to exceed the perfect form boundary at MMC. When this is done, the note "PERFECT FORM AT MMC NOT REQUIRED" must accompany the size dimension.

Perfect form may also be violated when the straightness of a cylindrical object is specified in relation to the axis. (This is discussed later in this chapter.) This is where the feature control frame is displayed with the diameter dimension and a diameter symbol precedes the geometric tolerance.

REGARDLESS OF FEATURE SIZE (RFS)

Regardless of feature size is the term used to indicate that a geometric tolerance or datum reference applies at any increment of size of the feature within its size tolerance.

The following rules govern the use of RFS:

❏ ASME Y14.5M "Rule 2–All Applicable Geometric Tolerances" states that RFS applies, with respect to the individual geometric tolerance and/or datum reference, when no material condition symbol is specified. Refer back to Example 4-2 for reference. MMC or LMC must be specified in the feature control frame where it is required.

❏ The geometric tolerances of circular runout, total runout, concentricity, and symmetry are applied *only* on an RFS basis. An MMC or LMC material condition symbol may not be used with these geometric characteristics.

❏ The geometric tolerance specified using RFS is held at any produced size within the specified dimensional tolerance.

❏ A datum reference specified with RFS means that the feature is centered about the axis or center plane regardless of the feature size.

As you learn to apply RFS to different drawing presentations throughout the rest of this text, keep the following ideas in mind:

❏ RFS is assumed for all geometric tolerances and datum references unless otherwise specified to be MMC or LMC.

❏ Regardless of feature size, as the name implies, means that the stated geometric tolerance is applied the same at any produced size–*regardless of feature size.*

Surface Control, Regardless of Feature Size (RFS)

Surface geometric control is when the feature control frame is either connected with a leader to the surface of the object or feature, or extended from an extension line from the surface of the object or feature. The use of a leader connecting the feature control frame to the surface is shown in Example 4-6.

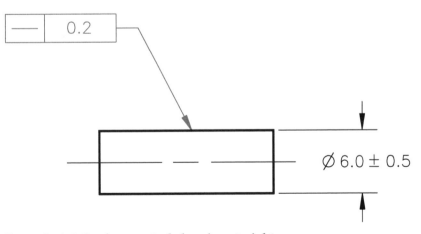

Example 4-6. Surface control showing straightness.

RFS is implied in the drawing in Example 4-6 because neither MMC or LMC is specified in the feature control frame. The surface straightness specification, as shown in Example 4-6, means that each longitudinal element of the surface must lie between two parallel lines, 0.2 apart, where the two lines and the nominal axis of the part share a common plane. Also, the feature must be within the specified size limits and within the perfect form boundary at MMC. When the actual size of the feature is MMC, then zero geometric tolerance is required. When the actual produced size departs from MMC, then the geometric tolerance is allowed to increase equal to the amount of departure until the specified geometric tolerance is reached. When the geometric tolerance specified in the feature control frame is reached, then the geometric tolerance stays the same at every other produced size. Example 4-7 shows an analysis of regardless of feature size (RFS) surface control based on the drawing presented in Example 4-6.

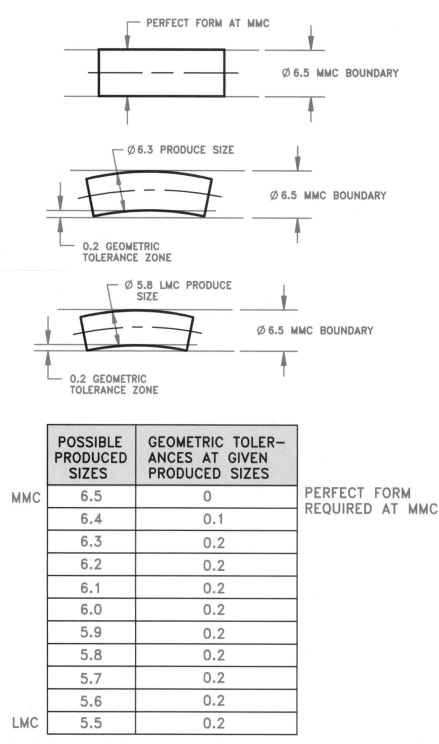

	POSSIBLE PRODUCED SIZES	GEOMETRIC TOLER- ANCES AT GIVEN PRODUCED SIZES	
MMC	6.5	0	PERFECT FORM REQUIRED AT MMC
	6.4	0.1	
	6.3	0.2	
	6.2	0.2	
	6.1	0.2	
	6.0	0.2	
	5.9	0.2	
	5.8	0.2	
	5.7	0.2	
	5.6	0.2	
LMC	5.5	0.2	

Example 4-7. The effect of specifying surface straightness. RFS is assumed and perfect form is required at MMC.

Axis Control, Regardless of Feature Size (RFS)

Axis geometric control is implemented by placing the feature control frame with the diameter dimension of the related object or feature. A good way to remember the difference between surface and axis control is to recognize that surface control is when the feature control frame is connected to the surface by a leader or extension line, while axis control places the feature control frame with the diameter dimension that correlates with the axis. When axis control is used, a diameter tolerance zone must be specified by placing the diameter symbol in front of the geometric tolerance in the feature control frame, as shown in Example 4-8.

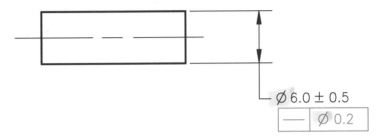

Example 4-8. Axis control showing straightness.

When axis control is specified, the perfect form boundary may be violated. This violation is permissible when the feature control frame is associated with the size dimension. When MMC is not specified, then RFS is implied. When this situation occurs, the geometric tolerance at various produced sizes remains the same as given in the feature control frame; even at MMC. Example 4-9 shows an analysis of axis control at regardless of feature size (RFS) based on the drawing in Example 4-8.

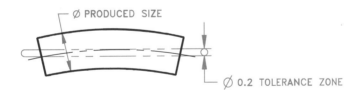

	POSSIBLE PRODUCED SIZES	GEOMETRIC TOLERANCES AT GIVEN PRODUCED SIZES	
MMC	6.5	0.2	PERFECT FORM NOT REQUIRED AT MMC
	6.4	0.2	
	6.3	0.2	
	6.2	0.2	
	6.1	0.2	
	6.0	0.2	
	5.9	0.2	
	5.8	0.2	
	5.7	0.2	
	5.6	0.2	
LMC	5.5	0.2	

Example 4-9. The effect of specifying axis straightness. RFS is assumed and perfect form is *not* required at MMC.

MAXIMUM MATERIAL CONDITION (MMC) Ⓜ

Maximum material condition is the condition where a feature contains the maximum amount of material within the stated limits of size. For example, maximum shaft diameter or minimum hole diameter are both MMC. When MMC is used in the feature control frame, the given geometric tolerance is maintained when the feature is produced at MMC. Then as the actual produced size departs from MMC, the geometric tolerance is allowed to get larger equal to the amount of departure from MMC. Think of using the MMC material condition symbol as meaning "at MMC." In other words, the specified geometric tolerance or datum reference is held only at the MMC produced size.

One of the following formulas can be used to calculate the geometric tolerance at any produced size when a MMC material condition symbol is used. The formula you should use is determined by whether the part is an external feature, such as a shaft, or an internal feature, such as a hole.

External Feature:
    MMC – Produced Size + Given Geometric Tolerance = Applied Geometric Tolerance

Internal Feature:
    Produced Size – MMC + Given Geometric Tolerance = Applied Geometric Tolerance

Axis Control, Maximum Material Condition (MMC)

When it is desirable to use MMC as the material condition symbol, then the MMC symbol must be placed in the feature control frame after the geometric tolerance or datum reference. This axis control is also a diameter tolerance zone and the diameter symbol must precede the geometric tolerance, as shown in Example 4-10.

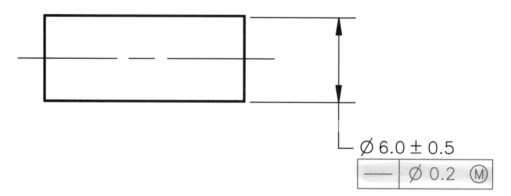

$\emptyset\, 6.0 \pm 0.5$

| — | Ⓜ $\emptyset\, 0.2$ Ⓜ |

Example 4-10. Axis control showing straightness with the MMC material condition symbol applied.

When a MMC material condition symbol is used, the geometric tolerance is the same as specified in the feature control frame at the MMC produced size. Then, as the produced size departs from MMC, the geometric tolerance is allowed to increase equal to the amount of departure from MMC. For example, the geometric tolerance at MMC in Example 4-10 is 0.2. If the part is produced at MMC, the geometric tolerance is 0.2. If the part is produced at 6.1, the applied geometric tolerance is: MMC (6.5) – PRODUCED SIZE (6.1) + GIVEN GEOMETRIC TOLERANCE (0.2) = APPLIED GEOMETRIC TOLERANCE (0.6). The maximum geometric tolerance is at the LMC produced size. LMC is the condition where a feature of size contains the least amount of material within the limits. The complete interpretation of the drawing shown in Example 4-10 is given in Example 4-11.

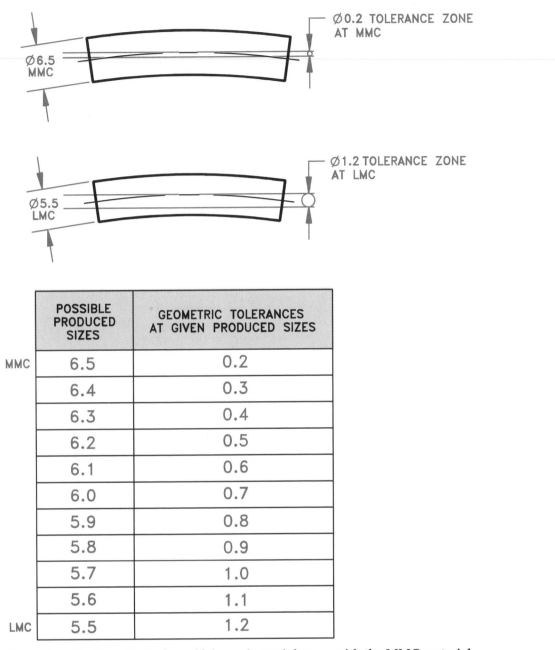

POSSIBLE PRODUCED SIZES	GEOMETRIC TOLERANCES AT GIVEN PRODUCED SIZES
6.5	0.2
6.4	0.3
6.3	0.4
6.2	0.5
6.1	0.6
6.0	0.7
5.9	0.8
5.8	0.9
5.7	1.0
5.6	1.1
5.5	1.2

MMC (6.5 row) ... LMC (5.5 row)

Example 4-11. The effect of specifying axis straightness with the MMC material condition symbol used.

The concepts of surface and axis straightness previously discussed may also be applied on an RFS or MMC basis to noncylindrical features of size. When this is done, the derived center plane must lie in a geometric tolerance zone between two parallel planes separated by the amount of the dimensional tolerance. Otherwise, the feature control frame placement is the same as previously discussed. The diameter tolerance zone symbol is not used in front of the geometric tolerance because the tolerance zone is noncylindrical, established by two parallel planes as shown in Example 4-12.

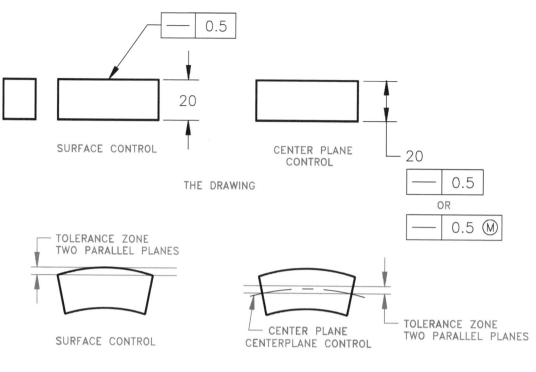

Example 4-12. Surface and center plane controls for noncylindrical features.

LEAST MATERIAL CONDITION (LMC) Ⓛ

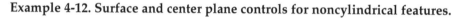

Least material condition is the condition where the feature of size contains the least amount of material. For example, minimum shaft diameter or maximum hole diameter are both conditions of LMC. LMC is the opposite of MMC. When an LMC material condition symbol is used in the feature control frame, the given geometric tolerance is held at the LMC produced size. When the actual produced size departs from LMC toward MMC, the geometric tolerance is allowed to increase equal to the amount of departure. The maximum geometric tolerance is at the MMC produced size. The formula for calculating the applied geometric tolerance in an LMC application is based on the relationship to an external or internal feature as follows:

External Feature

Produced Size – LMC + Given Geometric Tolerance = Applied Geometric Tolerance

Internal Feature

LMC – Produced Size + Given Geometric Tolerance = Applied Geometric Tolerance

Example 4-13 shows an application of LMC in the feature control frame where the axis perpendicularity of a hole must be within a 0.2 diameter geometric tolerance zone, at LMC, to Datum A. When the feature size is at LMC (12.5) the geometric tolerance is held as specified in the feature control frame. As the actual produced size decreases toward MMC, the geometric tolerance increases equal to the amount of change from LMC to the maximum change at MMC. The analysis in Example 4-13 shows the possible geometric tolerance variation from LMC to MMC. This specification is often used to control the minimum wall thickness of the part.

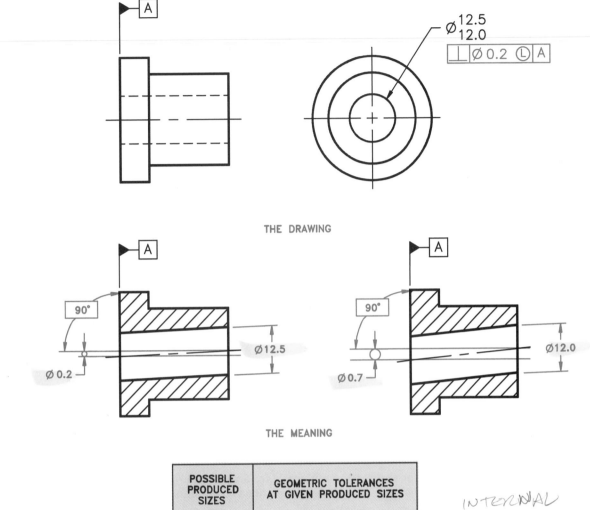

Example 4-13. The effect of specifying axis perpendicularity with the LMC material condition symbol used.

Perfect Form Boundary, LMC

There is no requirement for the feature to maintain perfect form when produced at its LMC limit of size when an LMC material condition symbol is specified. The feature is permitted to vary from true form to the maximum variation allowed by the perfect form boundary at MMC.

PRIMARY DATUM FEATURE, RFS

Datum features such as diameters and widths that are influenced by size variations are also subject to variations in form. RFS is implied in these cases unless otherwise specified. When a datum feature has a size dimension and a geometric form tolerance, the size of the simulated datum is the MMC size limit. This applies, except for axis straightness where the boundary is allowed to exceed MMC.

When a datum feature of size is represented on an RFS basis, the datum is established by contact between the datum feature surface and the surface of processing equipment such as a centering device. The processing equipment establishes the datum axis or center plane because it acts as the true geometric match or counterpart of the datum feature.

When a datum axis is primary and applied at RFS, the simulated datum is the axis of the processing equipment called the *true geometric counterpart.* The true geometric counterpart for an external feature is the smallest circumscribed perfect cylinder that contacts the datum feature surface, as shown in Example 4-14. Imagine this as placing a cylinder around the datum feature until the cylinder closes in on, and touches, the high points of the datum feature.

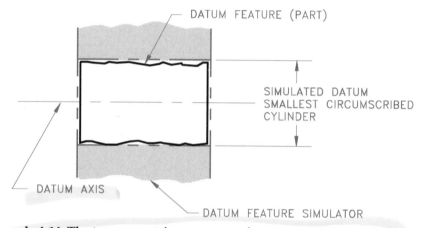

Example 4-14. The true geometric counterpart for an external feature is the smallest circumscribed perfect cylinder that contacts the datum feature surface.

The true geometric counterpart for an internal feature is the largest inscribed perfect cylinder that contacts the datum feature surface, as shown in Example 4-15. Imagine this as placing a cylinder inside the datum feature and increasing the size of the cylinder until the cylinder touches the high points of the datum feature.

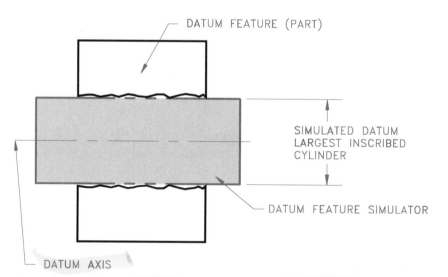

Example 4-15. The true geometric counterpart for an internal feature is the largest inscribed perfect cylinder that contacts the datum feature surface.

When a datum center plane is primary and applied at RFS, the simulated datum is the center plane of the true geometric counterpart. The true geometric counterpart for an external feature is two parallel planes that contact the datum feature surface at minimum separation, as shown in Example 4-16. Imagine this as closing vice jaws down on the datum surface until the vice jaws touch the high points of the datum feature.

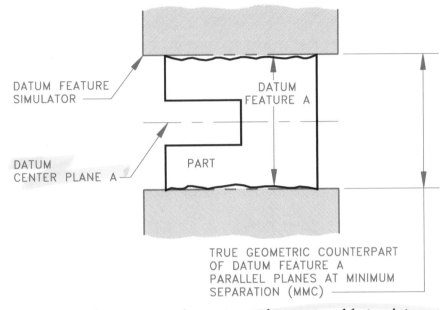

Example 4-16. The true geometric counterpart for an external feature is two parallel planes at minimum separation that contacts the datum feature surface.

The true geometric counterpart for an internal feature is two parallel planes at maximum separation as shown in Example 4-17. Imagine this as opening two parallel plates until the plates touch the high points of the datum feature.

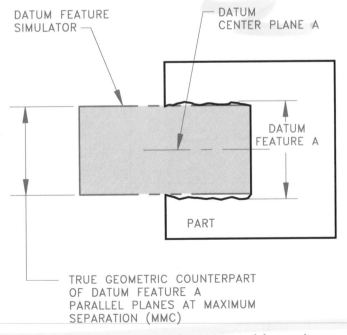

Example 4-17. The true geometric counterpart for an internal feature is two parallel planes at maximum separation that contacts the datum feature surface.

SECONDARY AND TERTIARY DATUM FEATURE, RFS

The secondary datum for either an axis or center plane is established the same way as previously discussed for the primary datum axis or center plane with the following additional requirement:

❑   The contacting cylinder or parallel planes of the true geometric counterpart must be 90°, or another design angle, to the primary datum. The primary datum is usually an adjacent plane. Refer to Example 4-18.

The tertiary datum for either an axis or center plane is established the same way as just discussed for the secondary datum axis or center plane with the following additional requirement:

❑   The contacting cylinder or parallel planes of the true geometric counterpart must be 90°, or another design angle, to both the primary and secondary datums. The tertiary datum feature may be oriented with a datum axis, as shown in Example 4-18, or offset from a plane of the datum reference frame.

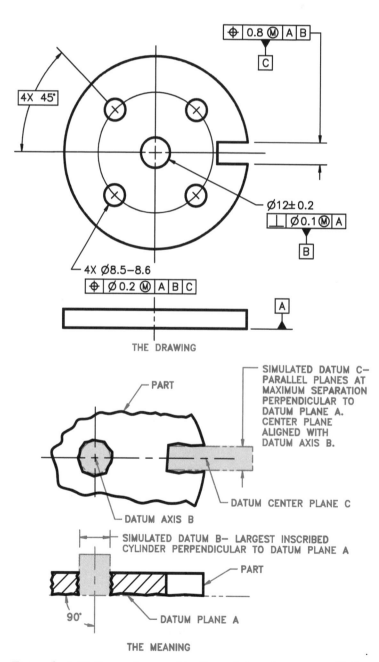

Example 4-18. Secondary and tertiary datum features at RFS.

DATUM PRECEDENCE AND MATERIAL CONDITION

The effect of material condition on the datum and related feature may be altered by changing the datum precedence and the applied material condition symbol. The *datum precedence* is established by the order of placement in the feature control frame. The first datum listed is the primary datum, followed by the secondary and tertiary datums. Refer to Example 4-19.

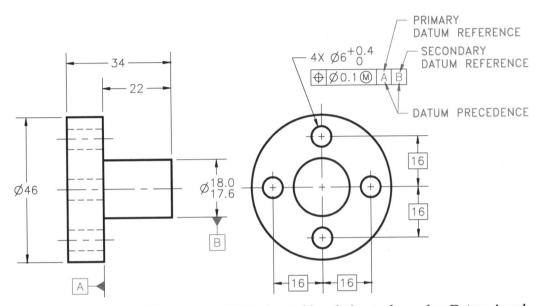

Example 4-19. A part with a pattern of holes located in relation to the surface Datum A and the ∅18.0/17.6 axis Datum B.

Change in size and form is allowed by the size tolerance of the datum feature. It is important to determine the datum precedence and material condition since these changes alter the fit and design function of the part.

The drawing in Example 4-19 shows a part with a pattern of holes located in relation to the surface Datum A and the surface Datum B. The datum requirements for the position tolerance associated with the location of the 4 holes may be specified in three different ways, as follows.

The illustration in Example 4-20 shows the surface Datum A as primary and the axis Datum B as secondary. Datum Plane A is established first, followed by Datum Axis B. Datum Axis B is established by the true geometric counterpart of Datum Feature B, which is the smallest circumscribed cylinder perpendicular to Datum Plane A.

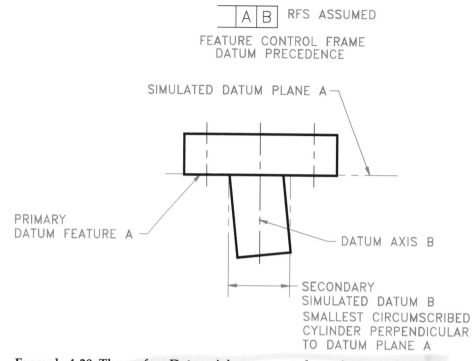

Example 4-20. The surface Datum A is represented as primary and the axis Datum B is secondary.

The illustration in Example 4-21 shows the Datum Axis B as primary and the Datum Surface A as secondary. Datum Axis B is established first followed by Datum Surface A. Datum Axis B is established by the true geometric counterpart of Datum Feature B, which is the smallest circumscribed cylinder. Datum Plane A is perpendicular to Datum Axis B.

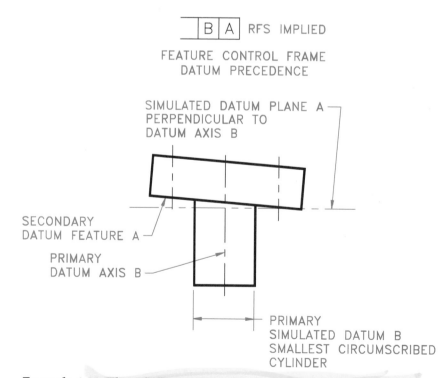

Example 4-21. The axis Datum B is shown as primary and the surface Datum A is secondary.

The illustration in Example 4-22 shows the surface Datum A as primary and the axis Datum B as secondary, with a MMC material condition. Datum Plane A is established first, followed by Datum Axis B. Datum Axis B is established by the true geometric counterpart of Datum Feature B, which is a cylinder equal in diameter to virtual condition and perpendicular to Datum Plane A. ***Virtual condition*** is the combined maximum material condition and geometric tolerance (Virtual Condition = MMC + Geometric Tolerance). Virtual condition is discussed in detail in Chapter 8.

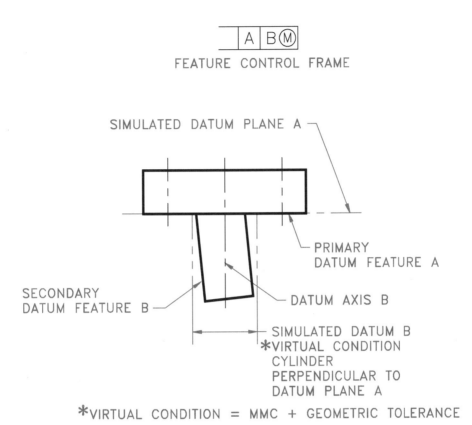

Example 4-22. The surface Datum A is represented as primary and the axis Datum B is secondary with an MMC material condition.

Name ______________________________

Test 4
Material Condition Symbols

1. Define perfect form boundary. ______________________________

____

____

____

____

____

2. Define regardless of feature size (RFS). ______________________________

____

____

____

____

____

3. ______________________________ is assumed for all geometric tolerance applications and related datum references unless otherwise specified.

4. How is a feature control frame connected to a related feature when surface control is intended? ______________________________

____

____

____

____

____

____

5. May perfect form at MMC be violated for surface straightness? Yes or No.

6. Given the following drawing and a list of possible produced sizes, specify the geometric tolerance at each possible produced size.

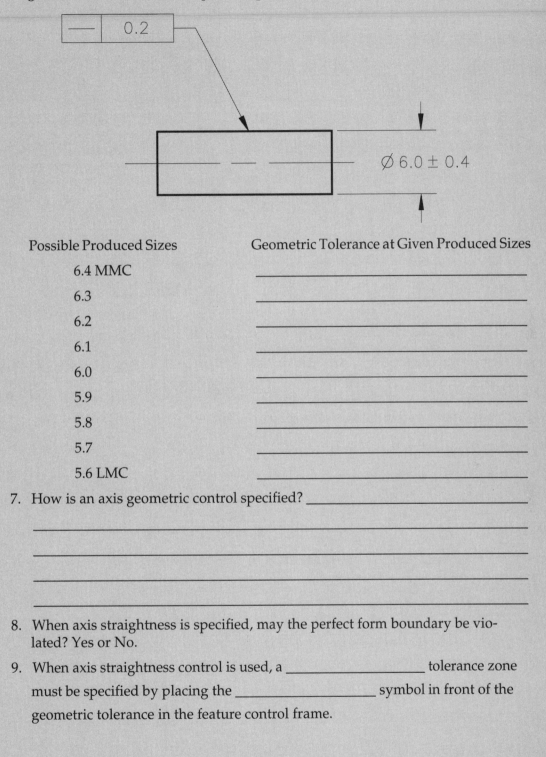

Possible Produced Sizes	Geometric Tolerance at Given Produced Sizes
6.4 MMC	
6.3	
6.2	
6.1	
6.0	
5.9	
5.8	
5.7	
5.6 LMC	

7. How is an axis geometric control specified? _____
_____
_____
_____
_____

8. When axis straightness is specified, may the perfect form boundary be violated? Yes or No.

9. When axis straightness control is used, a _____ tolerance zone must be specified by placing the _____ symbol in front of the geometric tolerance in the feature control frame.

10. Given the following drawing and a list of possible produced sizes, specify the geometric tolerance at each possible produced size.

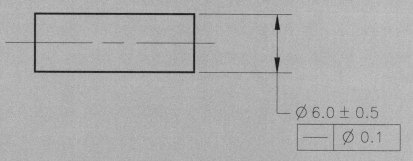

Ø 6.0 ± 0.5

| — | Ø 0.1 |

Possible Produced Sizes	Ø Geometric Tolerance at Given Produced Sizes
6.5 MMC	_____
6.4	_____
6.2	_____
6.0	_____
5.8	_____
5.6 LMC	_____
5.5 LMC	_____

11. Give the formula that can be used for calculating the geometric tolerance of an external feature at a given produced size when MMC is specified with the geometric tolerance in the feature control frame.

_____

_____

_____

12. Given the following drawing and a list of possible produced sizes, specify the geometric tolerance at each possible produced size.

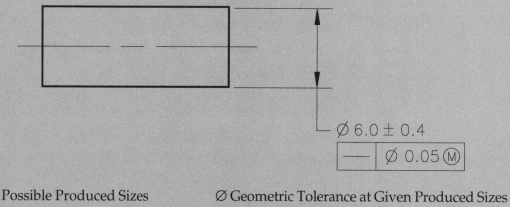

Ø 6.0 ± 0.4

| — | Ø 0.05 Ⓜ |

Possible Produced Sizes	Ø Geometric Tolerance at Given Produced Sizes
6.4 MMC	_____
6.2	_____
6.0	_____
5.8	_____
5.6 LMC	_____

13. Give the formula that may be used for calculating the geometric tolerance of an internal feature at a given produced size when LMC is specified with the geometric tolerance in the feature control frame. _____

   _____

   _____

14. The use of the LMC material condition symbol after the geometric tolerance in a feature control is often used to control minimum wall thickness. True or False?

15. When a datum feature has a size dimension and a geometric form tolerance, the size of the simulated datum is the _____ size limit, except for _____ straightness applications where the boundary is allowed to exceed MMC.

16. The effect of material condition on the datum and related feature may be altered by changing the datum precedence and the applied material condition symbol. True or False?

17. Given the following drawing and a list of possible produced sizes, specify the geometric tolerance at each possible produced size.

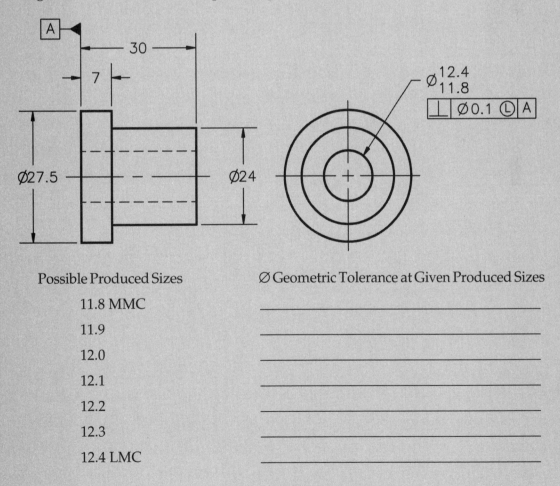

Possible Produced Sizes	∅ Geometric Tolerance at Given Produced Sizes
11.8 MMC	_____
11.9	_____
12.0	_____
12.1	_____
12.2	_____
12.3	_____
12.4 LMC	_____

18. Which of the following statements are true in regards to datum precedence and datum reference? (More than one may be correct.)

   a) Datum precedence is established by the order of placement in the feature control frame.

   b) Datum precedence is established by alphabetical order of datum reference letters.

   c) The first datum listed in the feature control frame is the primary datum reference.

   d) "A" is always the primary datum.

   e) The third datum listed in the feature control frame is the tertiary datum reference.

   f) RFS is assumed unless otherwise specified.

Print Reading Exercise

Name: _____

4

The following print reading exercise uses actual industry prints with related questions that require you to read specific dimensioning and tolerancing representations. The answers should be based on the previously learned content of this book. The prints used are based on ASME standards. However, company standards may differ slightly. When reading these prints, or any other industry prints, a degree of flexibility may be required to determine how individual applications correlate with the ASME standards.

Print Reading Exercise

Refer to the print of the SLEEVE-DEWAR REIMAGING found on page 303.

1. Refer to the ∅.8740±.0005 dimension:

   a) What is the MMC? _____

   b) What is the LMC? _____

   c) Name the geometric characteristic symbol found in the feature control frame.

   _____

   d) What is the geometric tolerance? _____

   e) Is this geometric tolerance an axis or surface control? _____

   f) Given the following list of possible produced sizes, determine the geometric tolerance at each produced size:

Possible Sizes	Geometric Tolerance
.8745	_____
.8742	_____
.8740	_____
.8738	_____
.8735	_____

2. Refer to the ∅.674 dimension:

   a) What is the tolerance? _____

   b) What is the MMC? _____

   c) What is the LMC? _____

   d) Name the geometric characteristic symbol found in the feature control frame.

   _____

   e) What is the geometric tolerance? _____

   f) What is the material condition symbol associated with this geometric

   tolerance? _____

   g) Is this geometric tolerance an axis or surface control? _____

h) Give the relationship of the datum reference to the geometric tolerance.____

_____

_____

_____

_____

i) Given the following list of possible produced sizes, determine the geometric tolerance at each produced size:

Possible Sizes	Geometric Tolerance
.677	_____
.676	_____
.675	_____
.674	_____
.673	_____

3. Refer to the ⌀.750 dimension:

a) What is the tolerance?_____

b) What is the MMC?_____

c) What is the LMC? _____

d) Name the geometric characteristic symbol found in the feature control frame.

_____

e) What is the geometric tolerance? _____

f) What is the material condition symbol associated with this geometric tolerance? _____

g) Is this geometric tolerance an axis or surface control? _____

h) Given the following list of possible produced sizes, determine the geometric tolerance at each produced size:

Possible Sizes	Geometric Tolerance
.753	_____
.752	_____
.751	_____
.750	_____
.749	_____

Refer to the print of the PEDAL-ACCELERATOR found on page 306.

4. Refer to the 19.10-19.08 diameter dimension at Datum D and answer the following related questions:

a) What is the MMC?_____

b) What is the LMC? _____

c) The position tolerance for this feature is referenced to which datums? _____

d) Why is there no material condition symbol after the perpendicularity geometric tolerance? _____

_____

_____

_____

e) Given the following list of possible produced sizes, determine the geometric tolerance at each produced size for perpendicularity and position:

Possible Sizes	Geometric Tolerance	
	Perpendicularity	Position
19.100	_____	_____
19.095	_____	_____
19.090	_____	_____
19.085	_____	_____
19.080	_____	_____

Refer to the print of the MOUNTING PLATE (UPPER)-FRAME ASSY 3 AXIS HP found on page 307.

5. Refer to the feature control frame with Datum A.

a) What is the geometric characteristic symbol associated with this feature control frame? _____

b) What is the geometric tolerance? _____

c) How do any changes in produced sizes effect the geometric tolerance? ____

_____

_____

_____

6. Refer to the .2510±.0005 dimension:

a) What is the tolerance? _____

b) What is the MMC? _____

c) What is the LMC? _____

d) Name the geometric characteristic symbols found in the feature control frames. _____

e) Give the geometric tolerances with each geometric characteristic. _____

_____

f) What is the material condition symbol associated with each geometric tolerance? _____

_____

g) Do the geometric tolerances apply to an axis, center plane, or surface control?

_____

h) Given the following list of possible produced sizes, determine the geometric tolerance at each produced size:

Possible Sizes	Geometric Tolerance	
	Parallelism	Position
.2505	_____	_____
.2510	_____	_____
.2515	_____	_____

Refer to the print of the HYDRAULIC VALVE found on page 308.

7. Refer to the ∅.344 dimension:

a) What is the tolerance?_____

b) What is the MMC?_____

c) What is the LMC? _____

d) Name the geometric characteristic symbol found in the feature control frame.

_____

e) What is the geometric tolerance? _____

f) What is the material condition symbol associated with this geometric tolerance? _____

g) Given the following list of possible produced sizes, determine the geometric tolerance at each produced size:

Possible Sizes	Geometric Tolerance
.349	_____
.347	_____
.345	_____
.343	_____
.341	_____
.339	_____

8. Refer to DIM 'A' part number 1 MS 2427-3:

a) What is the tolerance?_____

b) What is the MMC?_____

c) What is the LMC? _____

d) Name the geometric characteristic symbol found in the feature control frame.

_____

e) What is the geometric tolerance? _____

f) What is the material condition symbol associated with this geometric tolerance? _____

g) Is this geometric tolerance an axis or surface control?_____

h) Given the following list of possible produced sizes, determine the geometric tolerance at each produced size:

Possible Sizes	Geometric Tolerance
.687	_____
.686	_____
.685	_____
.684	_____

Refer to the print of the HOUSING-LENS, FOCUS found on page 313.

9. Refer to the Ø.122 THRU COUNTERBORE Ø.188 DEPTH .15 dimension: (Note: the geometric tolerance is applied the same to the hole and counterbore)

a) What is the MMC of the hole? _____

b) What is the LMC of the hole? _____

c) What is the MMC of the counterbore? _____

d) What is the LMC of the counterbore? _____

e) Given the following list of possible produced sizes, determine the geometric tolerance at each produced size:

Possible Sizes–Hole	Geometric Tolerance
.125	_____
.124	_____
.123	_____
.122	_____
.121	_____

Possible Sizes–Counterbore	Geometric Tolerance
.191	_____
.190	_____
.189	_____
.188	_____
.187	_____

Precision machining is often performed on computer numerical control (CNC) machines. Geometric dimensioning and tolerancing drawings created on a CADD system can be downloaded directly to the CNC machine.   (Maho Machine Tool Corp.)

Chapter 5

Tolerances of Form
and Profile

This chapter explains the concepts and techniques of dimensioning and tolerancing to control the form and profile of geometric shapes. *Form tolerances* control:

❑ Straightness.
❑ Flatness.
❑ Circularity.
❑ Cylindricity.

Profile tolerances control the form and/or orientation of straight lines or surfaces, arcs, and irregular curves.

When size tolerances provided in conventional dimensioning do not provide sufficient control for the functional design and interchangeability of a product, then form and/or profile tolerances should be specified. As discussed in Chapter 4, size limits control a degree of form. The extent of this control should be evaluated before specifying geometric tolerances of form or profile.

STRAIGHTNESS TOLERANCE

Straightness is a condition where an element of a surface or an axis is in a straight line. Straightness is a form tolerance. The *straightness tolerance* specifies a zone that the required surface element or axis must lie in. Example 5-1 shows a detailed example of the straightness geometric characteristic symbol used in a feature control frame.

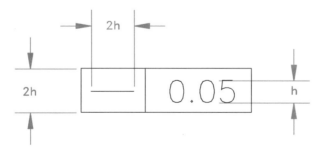

Example 5-1. Feature control frame with the straightness geometric characteristic symbol.

The *surface straightness tolerance* is represented by connecting the feature control frame to the surface with a leader, or by connecting the feature control frame to an extension line in the view where the surface to be controlled is shown as an edge. The feature may not exceed the MMC envelope and perfect form must be maintained if the actual size is produced at MMC. Otherwise, RFS applies and the geometric tolerance remains the same at any produced size. Example 5-2 shows a drawing and an exaggerated representation of what happens when a surface straightness tolerance is applied. Remember, straightness implies RFS. The chart in Example 5-2 shows the maximum out-of-straightness at several possible produced sizes. The straightness tolerance must be less than the size tolerance.

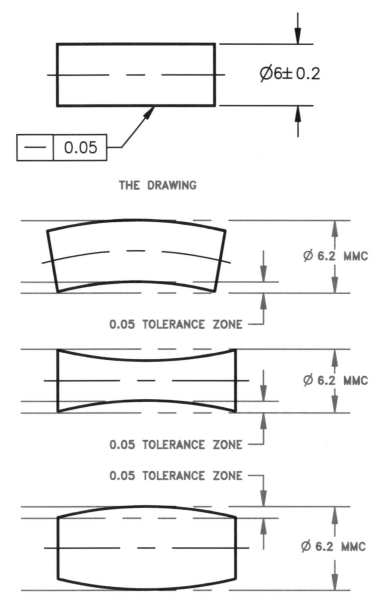

THE DRAWING

THE MEANING

	POSSIBLE PRODUCED SIZES	MAXIMUM OUT-OF-STRAIGHTNESS
MMC	6.2	* 0
	6.1	0.05
	6.0	0.05
	5.9	0.05
LMC	5.8	0.05

* PERFECT FORM REQUIRED

Example 5-2. The effect of surface straightness. RFS is assumed and perfect form is required at MMC.

Axis straightness is specified on a drawing by placing the feature control frame below the diameter dimension, and a diameter symbol is placed in front of the geometric tolerance to specify a cylindrical tolerance zone, as shown in Example 5-3. Notice in the chart in Example 5-3 that this application allows a violation of perfect form at MMC. RFS is assumed.

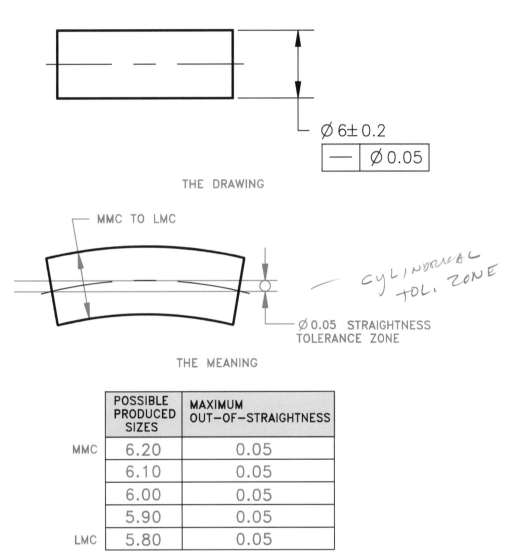

THE DRAWING

Ø 6± 0.2

THE MEANING

	POSSIBLE PRODUCED SIZES	MAXIMUM OUT–OF–STRAIGHTNESS
MMC	6.20	0.05
	6.10	0.05
	6.00	0.05
	5.90	0.05
LMC	5.80	0.05

Example 5-3. The effect of axis straightness. RFS is assumed and perfect form is *not* required at MMC.

Axis straightness may also be specified on an MMC basis by placing the MMC material condition symbol after the geometric tolerance. The specified geometric tolerance is then held at MMC and allowed to increase as the actual size departs from MMC. The geometric tolerance is at MMC, as shown in Example 5-4. In some cases the straightness tolerance may be greater than the size tolerance where necessary, but normally the straightness tolerance is less than the size tolerance.

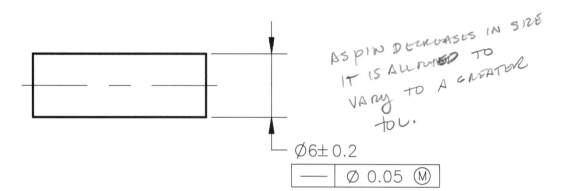

AS PIN DECREASES IN SIZE IT IS ALLOWED TO VARY TO A GREATER tol.

Ø6± 0.2

	POSSIBLE PRODUCED SIZES	MAXIMUM OUT—OF—STRAIGHTNESS
MMC	6.20	0.05
	6.10	0.15
	6.00	0.25
	5.90	0.35
LMC	5.80	0.45

Example 5-4. The effect of axis straightness with the MMC material condition symbol used.

Unit Straightness

Straightness per unit of measure may be applied to a part or feature in conjunction with a straightness specification over the total length. This may be done as a means of preventing an abrupt surface variation within a relatively short length of the feature. The specified geometric tolerance over the total length is greater than the unit tolerance and is normally given to keep the unit tolerance from getting out of control when applied to the length of the feature. The per unit specification may be given as a tolerance per inch or per 25 millimeters of length. When this technique is used, the feature control frame is doubled in height and split so the tolerance over the total length may be specified in the top half and the per unit control placed in the bottom half, as shown in Example 5-5. Caution should be exercised when using unit straightness without the limiting geometric tolerance over the total length, as this could cause excessive waviness in the feature or part.

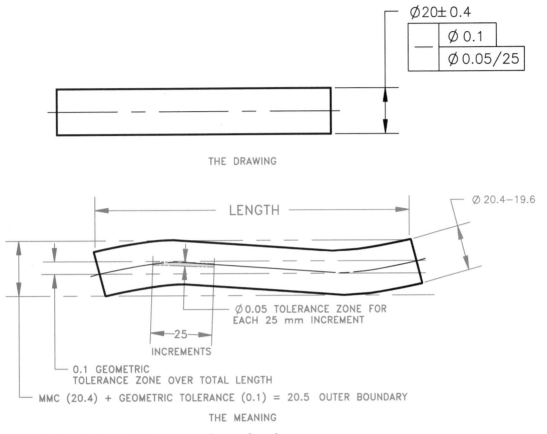

Example 5-5. Unit straightness used on a drawing.

In Example 5-5, the derived axis or centerline of the actual feature must lie within a cylindrical tolerance zone of 0.1 diameter for the total length and within a 0.05 cylindrical tolerance zone for any 25mm length, regardless of feature size. Additionally, each circular element of the surface must be within the specified limits of size.

Straightness of Noncylindrical Symmetrical Features

Straightness may also be applied on an RFS or MMC basis to noncylindrical features of size. When this is done, the associated center plane must lie within two parallel planes separated by a distance equal to the specified geometric tolerance zone. The feature control frame may be attached to the view where the surface appears as a line by using a leader or an extension line. In this situation, the diameter symbol is not placed in front of the geometric tolerance.

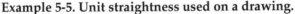

Straightness of a Flat Surface

Straightness may also be applied to a flat surface. When this is done, the straightness geometric tolerance may control single line elements on the surface in one or two directions. The direction of the tolerance zone is determined by the placement of the feature control frame, as shown in Example 5-6.

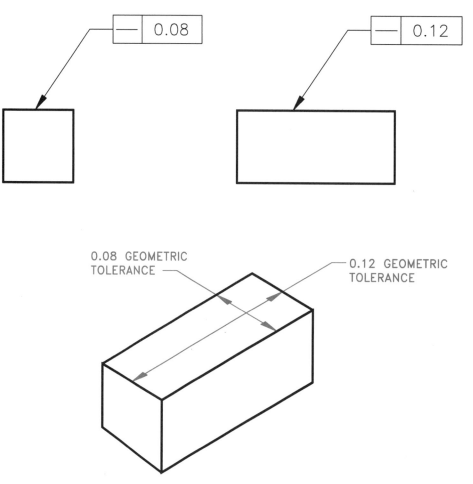

Example 5-6. Straightness applied to a flat surface.

FLATNESS TOLERANCE

Perfect flatness is the condition of a surface where all of the elements are in one plane. Flatness is a form tolerance. A *flatness tolerance zone* establishes the distance between two parallel planes that the surface must lie within. The flatness feature control frame is detailed in Example 5-7.

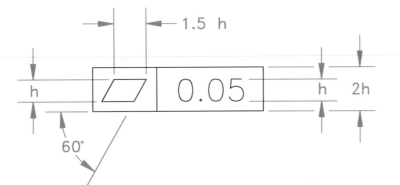

Example 5-7. Feature control frame with the flatness geometric characteristic symbol.

When a flatness geometric tolerance is specified, the feature control frame is connected by a leader or an extension line in the view where the surface appears as a line. Refer to Example 5-8. All of the points of the surface must be within the limits of the specified tolerance zone. The smaller the tolerance zone, the flatter the surface. The flatness tolerance must be less than the size tolerance when the surface is associated with a size tolerance.

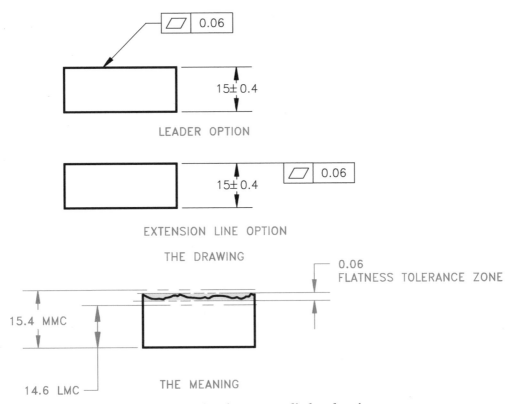

Example 5-8. The flatness geometric tolerance applied to drawings.

Specific Area Flatness

At times it may be necessary to provide a flatness callout for only a specific area of a surface. This procedure is known as *specific area flatness.* Specific area flatness may be used when a large cast surface must be flat but it is possible to finish only a relatively small area, rather than an expensive operation of machining the entire surface, as shown in Example 5-9.

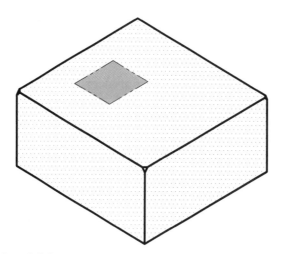

Example 5-9. A pictorial drawing showing specific area flatness on a large surface.

When specific area flatness is used, the specific area is outlined with phantom lines and then section lines are placed within the area. The specific area is then located, preferably from datums, with basic or ± dimensions. The feature control frame is then connected to the area with a leader line, as shown in Example 5-10.

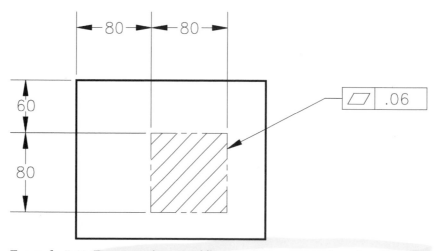

Example 5-10. Representing specific area flatness on a drawing. Basic or ± location dimensions may be used.

Unit Flatness

Unit flatness may be specified when it is desirable to control the flatness of a given surface area as a means of controlling an abrupt surface variation within a small area of the feature. The unit flatness specification may be used alone or in combination with a total tolerance. Most applications use unit flatness in combination with a total tolerance over the entire surface so the unit callout is not allowed to get out of control. When this is done, the height of the feature control frame is doubled with the total tolerance placed in the top half and the unit tolerance plus the size of the unit area placed in the bottom half. The unit tolerance must be smaller than the total tolerance, as shown in Example 5-11.

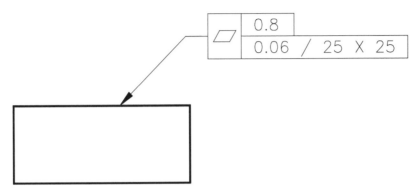

Example 5-11. Unit flatness specified on a drawing.

CIRCULARITY TOLERANCE

Circularity is characterized by any given cross section taken perpendicular to the axis of a cylinder or cone, or through the common center of a sphere. Circularity is a form tolerance. The *circularity geometric tolerance* is formed by a radius zone creating two concentric circles that the actual surface must lie within. The circularity feature control frame is detailed in Example 5-12. The tolerance applies to only one sectional element at a time, as shown exaggerated in Example 5-13.

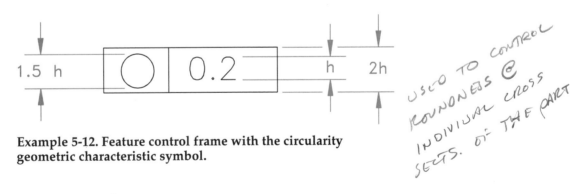

Example 5-12. Feature control frame with the circularity geometric characteristic symbol.

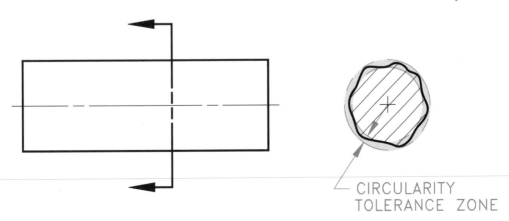

Example 5-13. An analysis of the circularity geometric tolerance.

The circularity feature control frame is connected with a leader to the view where the feature appears as a circle or in the longitudinal view, as shown in Example 5-14.

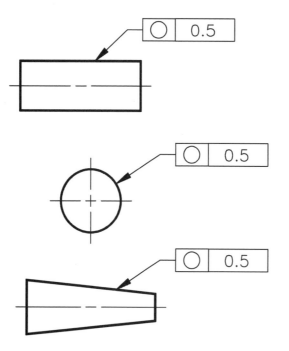

THE DRAWING OPTIONS

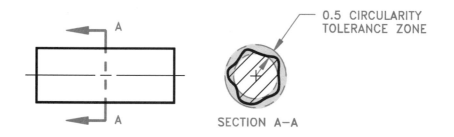

THE MEANING

Example 5-14. Applications of the circularity geometric tolerance.

Circularity Tolerance for a Sphere

The circularity geometric tolerance may also be applied to a sphere. When this is done, the circularity geometric tolerance is established by two concentric circles created by a plane passing through the center of the sphere. All points on the surface must lie within the circularity tolerance zone.

FREE STATE VARIATION

The circularity tolerance must be less than the size tolerance, except for parts subject to free state variation. *Free state variation* is the distortion of a part after removal of forces applied during manufacture. Distortion may happen to thin wall parts where the weight and flexibility of the part are effected by internal stresses applied during fabrication. These types of parts are referred to as *nonrigid.*

The part may have to meet the tolerance specifications while in free state or it may be necessary to hold features in a simulated mating part to verify dimensions. The free state symbol, shown in Example 5-15, is placed in the feature control frame after the geometric tolerance and material condition symbol (if MMC or LMC is required).

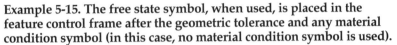

Example 5-15. The free state symbol, when used, is placed in the feature control frame after the geometric tolerance and any material condition symbol (in this case, no material condition symbol is used).

It may be necessary to specify circularity of a nonrigid part based on an average size diameter to help make sure that the desired diameter can be held during assembly. When this is necessary, the abbreviation "AVE" is placed after the size dimension, as shown in Example 5-16. The actual average diameter is the average of several

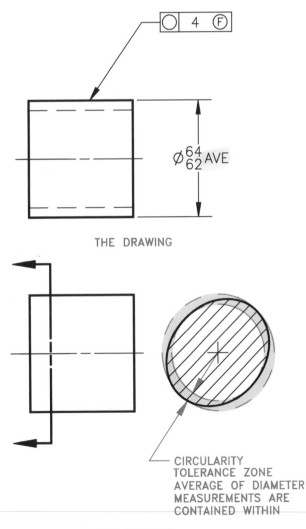

THE DRAWING

THE MEANING

Example 5-16. Specifying average size diameter of nonrigid parts using the "AVE" abbreviation after the size dimension.

measurements across the feature. The circularity tolerance is larger than the size tolerance, because the average of the diameter measurements is established within the boundaries of the circularity tolerance.

CYLINDRICITY TOLERANCE

Cylindricity is identified by a radius tolerance zone that establishes two perfectly concentric cylinders that the actual surface must lie within. Cylindricity is a form tolerance. The cylindricity feature control frame is detailed in Example 5-17.

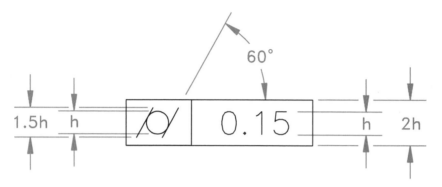

Example 5-17. Feature control frame with the cylindricity geometric characteristic symbol.

The feature control frame showing the cylindricity tolerance specification is connected by a leader to either the circular or longitudinal view, as shown in Example 5-18.

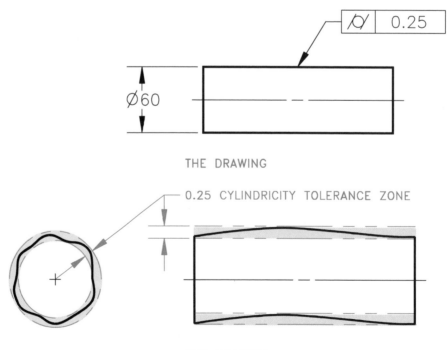

Example 5-18. Application of the cylindricity geometric tolerance.

Cylindricity can be characterized as a blanket tolerance that covers the entire feature. The difference between circularity and cylindricity is that:

❑ Circularity (circle) is a single cross sectional tolerance representing a zone between two concentric circles.

❑ Cylindricity (cylinder) is a tolerance that covers both circular and longitudinal elements at the same time representing a zone between two concentric cylinders.

PROFILE

Profile may be characterized as the outline of an object represented either by an external view or by a cross section through the object. The *true profile* or actual desired shape of the object is the basis of the profile tolerance. The true profile should be defined by basic dimensions in most applications. The *profile tolerance* specifies a uniform boundary along the true profile that the elements of the surface must lie within. Profile may be used to control form, or combinations of size, form, and orientation. When used as a refinement of size, the profile tolerance must be contained within the size tolerance.

MUST BE RELATED TO A DATUM

Profile tolerances are assumed to be bilateral. This means the tolerance zone is split equally on each side of the true profile. The profile tolerance zone may also be unilateral where the entire zone is on one side of the true profile. The profile tolerance may also be specified between two given points or all around the object. A profile tolerance is specified by connecting the feature control frame, using a leader, to the view or section that clearly shows the intended profile. There are two profile geometric characteristics: profile of a line and profile of a surface.

Profile of a Line

WHEN A PART HAS A VARIABLE CROSS SECTION

The *profile of a line tolerance* is a two dimensional or cross sectional geometric tolerance that extends along the length of the feature. The profile of a line symbol and associated feature control frame is shown in Example 5-19. The datum reference is provided in the feature control frame because the profile geometric tolerance zone is generally oriented to one or more datums. Profile of a line is used where it is not necessary to control the profile of the entire feature.

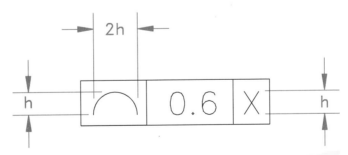

Example 5-19. Feature control frame with the profile of a line geometric characteristic symbol and datum reference.

Profile of a line tolerance is used in situations where parts or objects have changing cross sections throughout the length. An aircraft wing is an example of a part that has a changing cross section. Datums may be used in some situations, but are not necessary when the only requirement is the profile shape taken at various cross sections. When the leader from the profile feature control frame extends to the related surface without any additional clarification, the profile tolerance is assumed to be bilateral. Bilateral tolerances are equally split on each side of the basic dimensions that establish true profile.

Profile of a Line Between Two Points

The profile tolerance may be between two given points of the object. This specification is shown by using the "BETWEEN" symbol under the feature control frame. Any combination of letters may be used, such as "A" and "B" or "C" and "D." The true profile may be established by a basic or tolerance dimension. Profile of a line is a single cross sectional check anywhere along the intended surface. The profile tolerance is assumed to be *bilateral* unless otherwise specified. This means that the profile tolerance is split equally on each side of the true profile. The actual profile of the feature is confined within the profile tolerance zone. Refer to Example 5-20.

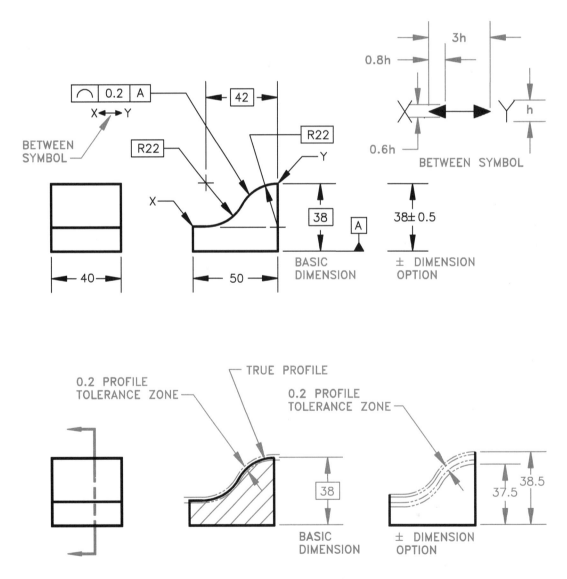

Example 5-20. Profile of a line specified between two points.

Profile of a Line All Around

Profile of a line may also specify a tolerance zone that goes around the entire object. When this is desired, the feature control frame is connected to the object with a leader as previously discussed and the "ALL AROUND" symbol is placed on the leader, as shown in Example 5-21. Any note indicating between two points is excluded.

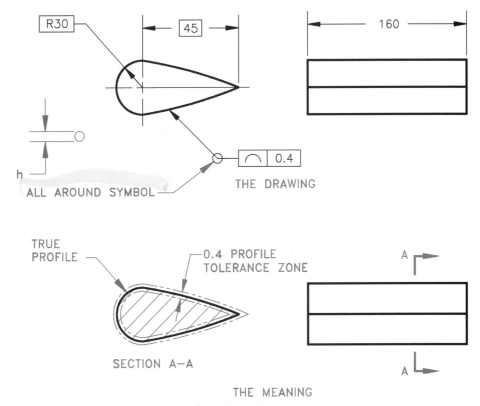

Example 5-21. Profile of a line specified all around.

Unilateral Profile of a Line

A bilateral profile tolerance is assumed unless unilateral specifications are provided. A ***unilateral profile*** is where the entire tolerance zone is on one side of the true profile. When a unilateral profile tolerance is required, a short phantom line is drawn parallel to the true profile on the side of the intended unilateral tolerance. A dimension line with arrowhead is placed on the far side and a leader line connects the feature control frame on the other side, as shown in Example 5-22.

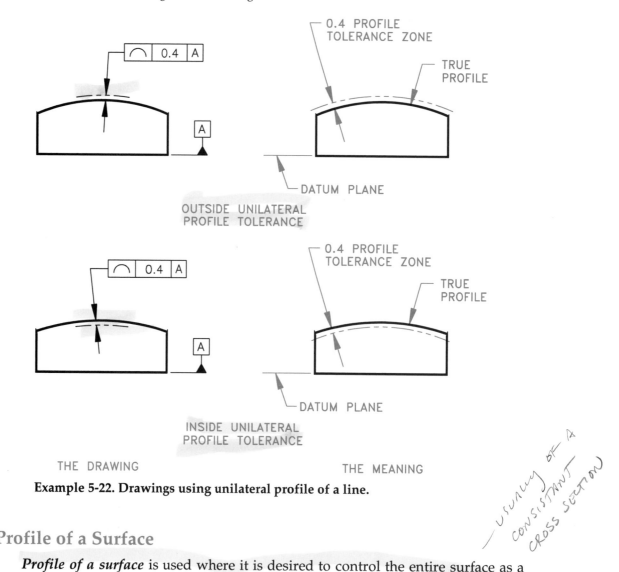

OUTSIDE UNILATERAL
PROFILE TOLERANCE

INSIDE UNILATERAL
PROFILE TOLERANCE

THE DRAWING   THE MEANING

Example 5-22. Drawings using unilateral profile of a line.

Profile of a Surface

Profile of a surface is used where it is desired to control the entire surface as a single entity. The profile of a surface geometric characteristic and associated feature control frame is detailed in Example 5-23. Profile of a surface is a blanket tolerance that is three dimensional extending along the total length and width or circumference of the object or feature(s). In most cases, the profile of a surface tolerance requires reference to datums for proper orientation of the profile.

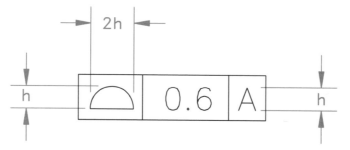

Example 5-23. Feature control frame with the profile of a surface geometric characteristic symbol and datum reference.

Profile of a Surface Between Two Points

The profile of a surface tolerance zone is bilateral unless otherwise specified, just as for profile of a line. Profile of a surface may be between two points and is handled in the same manner that was used for profile of a line between two points. A sample drawing and its related meaning is shown in Example 5-24.

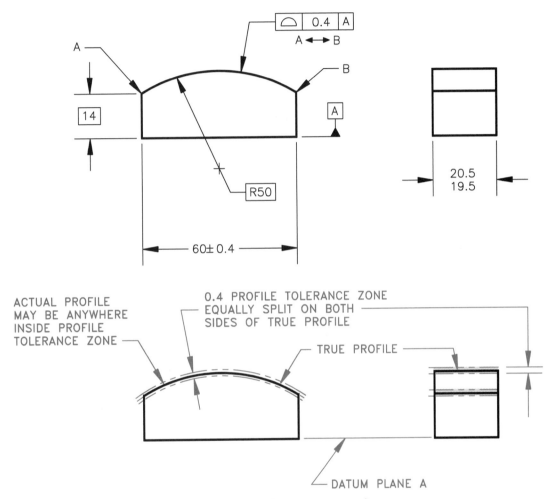

Example 5-24. Profile of a surface specified between two points.

Profile of a Surface All Around

Surface profile may also be applied to completely blanket objects that have a constant uniform cross section by placing the all around symbol on the leader line. When this is done, surfaces all around the object outline must lie between two parallel boundaries equal in width to the given geometric tolerance. The tolerance zone should also be perpendicular to a datum plane. Refer to Example 5-25.

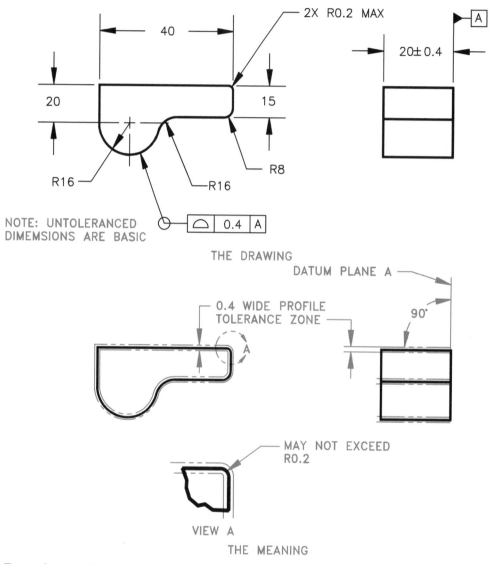

Example 5-25. Using profile of a surface all around.

Profile of a Sharp Corner

When a profile tolerance is at a sharp corner, the tolerance zone extends to the intersection of the boundary lines. In these situations, a rounded corner may occur, because the actual surface may be anywhere within the tolerance zone boundary. If this roundness must be controlled, then a maximum radius note shall be added to the drawing. The drawing in Example 5-25 has the note "R0.2 MAX" to indicate this.

Unilateral Profile of a Surface

Profile of a surface may also be specified as a unilateral tolerance by placing a short phantom line on the side of true profile where the unilateral tolerance zone is intended. The feature control frame is then connected to the phantom line with a leader, as shown in Example 5-26.

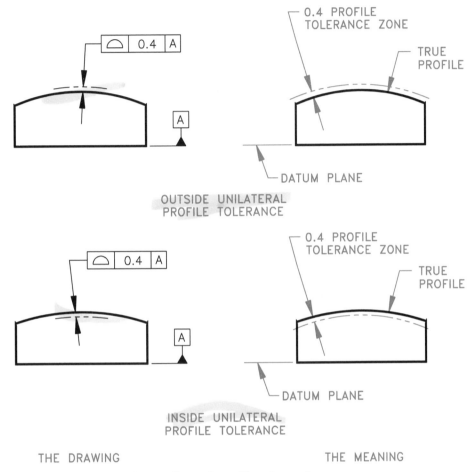

Example 5-26. Showing unilateral profile of a surface.

Profile of Coplanar Surfaces

Coplanar surfaces are two or more surfaces on a part that are on the same plane. A *coplanar profile tolerance* may be used when it is desirable to treat two or more separate surfaces, that lie on the same plane, as one surface. To control the profile of these surfaces as a single surface, place a phantom line between the surfaces in the view where the required surfaces appear as lines. Connect a leader from the feature control frame to the phantom line and add a note identifying the number of surfaces below the feature control frame. Refer to the note "2 SURFACES" shown in Example 5-27.

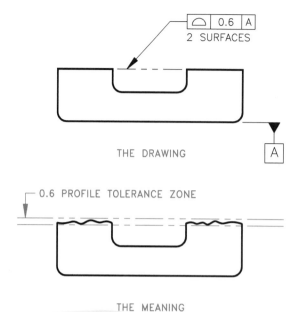

THE DRAWING

THE MEANING

Example 5-27. Profile of coplanar surfaces.

When there are several coplanar surfaces, it may be desirable to establish two surfaces as datum planes with a common profile tolerance such as the datum features labeled A and B in Example 5-28. Other coplanar surfaces may be controlled with a profile tolerance relative to both datums by placing the letters A-B in the feature control frame. The profile tolerance zone applies to all surfaces including the datums. Refer to Example 5-28.

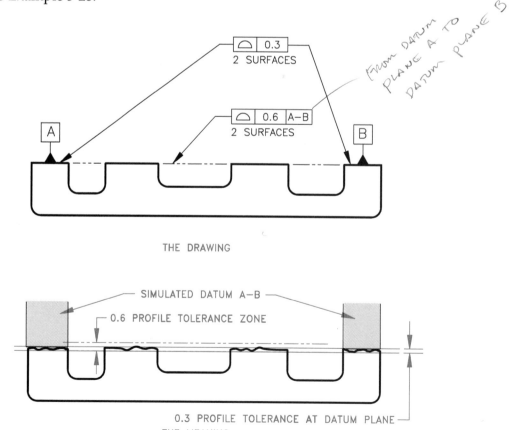

THE DRAWING

THE MEANING

Example 5-28. When there are several coplanar surfaces, it may be desirable to establish two surfaces as datum planes with a common profile tolerance and other surfaces controlled with a profile tolerance to both datums.

Profile of Plane Surfaces

Profile tolerancing may be used to control the form and orientation of plane surfaces. For example, profile of a surface may be used to control the angle of an inclined surface in relationship to a datum, as shown in Example 5-29. Notice in Example 5-29 that the required surface must lie between two parallel planes 0.1 apart equally split on each side of a true plane that has a basic angular orientation to a datum.

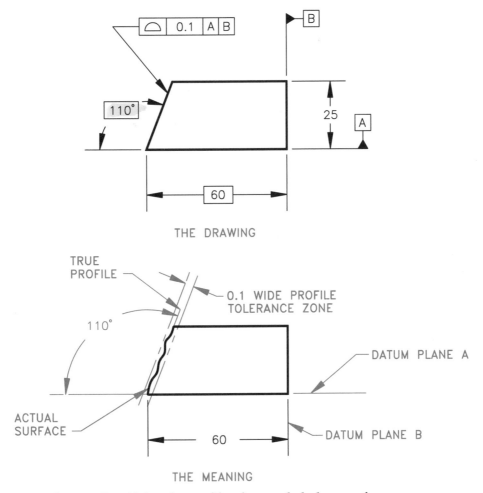

Example 5-29. Specifying the profile of an angled plane surface.

Profile of Conical Features

A profile tolerance may be used to control the form, or form and orientation, of a conical surface. The feature may be controlled independently as a refinement of size or may be oriented to a datum axis. In either case the profile tolerance must be within the size tolerance. Conical profile requires that the actual surface lie between two coaxial boundaries equal in width to the specified geometric tolerance, having a basic included angle, and within the size limits. (Coaxial means having the same axis.) Refer to Example 5-30.

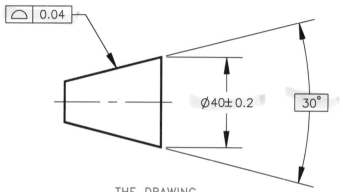

THE DRAWING

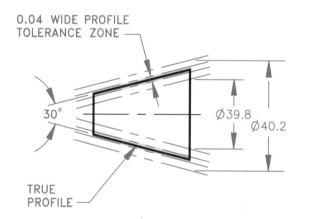

THE MEANING

Example 5-30. Specifying profile of a conical feature.

Composite Profile Tolerance

A *composite profile tolerance* provides for the location of a profiled feature and, at the same time, the control of form and orientation. This is done by doubling the height of the feature control frame that points to the feature to be controlled. The profile geometric characteristic symbol is placed in the first compartment. The top half of the feature control frame is called the *locating tolerance zone.* This is the profile tolerance that locates the feature from datums. The related datum reference is given in the order of precedence in the feature control frame and the feature to be controlled is located from datums with basic dimensions. The bottom half of the feature control frame is called the *profile form and orientation tolerance zone.* Datum referencing in the lower area establishes the limits of size, form, and orientation of the profile related to the locating tolerance zone. The actual feature surface must be within both tolerance zones, as shown in Example 5-31.

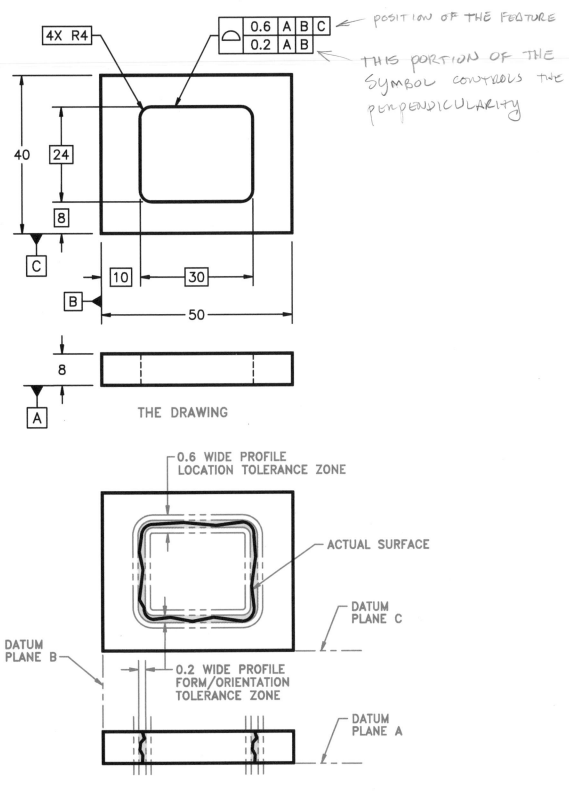

4X R4

⌒	0.6	A	B	C
	0.2	A	B	

← POSITION OF THE FEATURE

← THIS PORTION OF THE SYMBOL CONTROLS THE PERPENDICULARITY

40

24

8

C

10   30

B

50

8

A

THE DRAWING

0.6 WIDE PROFILE LOCATION TOLERANCE ZONE

ACTUAL SURFACE

DATUM PLANE C

DATUM PLANE B

0.2 WIDE PROFILE FORM/ORIENTATION TOLERANCE ZONE

DATUM PLANE A

THE MEANING

Example 5-31. A composite profile tolerance provides for the location of a profiled feature, and at the same time the control of form and orientation.

SPECIFYING BASIC DIMENSIONS IN A NOTE

To save time when an extensive number of basic dimensions are used on a drawing, a general note may be used to specify basic dimensions. Use the general note "UNTOLERANCED DIMENSIONS ARE BASIC" rather than using the customary rectangular block around the dimension to denote basic. Check with your company or instructor before using this method to determine if it is appropriate. Refer back to Example 5-25.

Test
Tolerances of Form and Profile

5

Name: _____

1. Name the geometric tolerance that specifies a zone that the required surface element or axis must lie within. ___FORM TOLERANCE___
   _____STRAIGHTNESS_____

2. Explain the difference between the methods used to represent surface and axis straightness. _____
   SURFACE - CONNECTED W/LEADER
   AXIS - DIRECTLY BELOW DIAMETER DIM.
   _____
   _____
   _____
   _____
   _____

3. Perfect form is required at MMC for ___SURFACE___ straightness.

4. The perfect form boundary may be violated at MMC for ___AXIS___ straightness.

5. Axis straightness may be specified on an MMC basis by placing the MMC symbol after the geometric tolerance in the feature control frame. The specified geometric tolerance is held at the MMC produced size. Explain what happens to the geometric tolerance as the produced size departs from MMC.
   ___THE GEOMETRIC TOLERANCE INCREASES___
   _____
   _____
   _____
   _____

6. Name the form geometric tolerance zone that establishes the distance between two parallel planes that the surface must lie within. ___FLATNESS___
   _____
   _____

7. Why should unit straightness have a limiting geometric tolerance over the total length? ___OTHERWISE EXCESSIVE WAVINESS___
   _____
   _____
   _____

8. What geometric tolerance is characterized by any given cross section taken perpendicular to the axis of a cylinder or cone, or through the common center of a sphere? _____

CIRCULARITY

9. Specify the appropriate views and the method of connecting a circularity feature control frame in a drawing. *CONN. W/LEADER TO VIEW*

_____

_____

_____

10. What is the difference between the circularity geometric tolerance and the cylindricity geometric tolerance? _____

CIRCULARITY - 2 DIMENSIONAL @ A SPECIFIC LOCAT.
CYLINDR. - 3 "

_____

_____

_____

_____

11. Which geometric tolerance, cylindricity or circularity, would require more precise control? *BECAUSE 3 DIMENSIONAL SPEC*

12. The *PROFILE* tolerance specifies a uniform boundary along the true profile that the elements of the surface must lie within.

13. Complete this statement: A profile tolerance is specified by connecting the feature control frame using a leader to *VIEW THAT DEPICTS PROFILE*

_____

_____

14. Name the two types of profile geometric characteristics. _____
PROFILE OF LINE OR PROFI. OF SURFACE

15. What situations, or types of features or parts, frequently require the use of a profile of a line tolerance? _____

_____

_____

16. How is a profile tolerance shown to be specified between two given points?
IDENT 2 PTS & USE BETWEEN

_____

_____

_____

_____

17. How is a profile tolerance specified all around an object or feature, rather than between two given points? _____ ALL AROUND _____

_____

_____

_____

18. Explain the difference between profile of a line and profile of a surface.

_____ LINE - 2 DIM, _____

_____ SURF - 3 DIM, _____

_____

_____

_____

_____

_____

19. Either the profile of a line or the profile of a surface may be all around, between two given points, unilateral, or bilateral. True or False?

20. Define a bilateral profile. _____ TOL ZONE EQUALY ON ± SIDES _____

_____

_____

_____

21. A bilateral profile is assumed unless otherwise specified. True or False?

22. Define a unilateral profile. _____

_____ ALL TOL. ZONE TO 1 SIDE _____

_____

_____

23. Explain how a unilateral profile tolerance zone is specified on a drawing.

_____ SHORT PHANTOM LINE DRAWN TO SHOW _____

_____ WHICH SIDE INTENDED _____

_____

_____

_____

24. Give the general note that may be used to specify basic dimensions that are used to dimension true profile rather than using the customary rectangular block around the dimension to indicate basic. This may be used to save time when an extensive number of basic dimensions are on the drawing.

_____

_____

_____

25. Define coplanar surfaces. __2 OR > SURF IN A PLANE__

26. Explain how to represent as a single surface the surface profile of four (4) coplanar surfaces. _____

    DRAW SKETCH   p 146

27. Profile tolerancing may be used to control the form and orientation of a plane surface. (True) or False?

28. Profile of a surface may be used to control the angle of an inclined surface in relationship to a datum. (True) or False?

29. A profile tolerance may be used to control the form, or form and orientation of a conical surface. (True) or False?

30. Define free state variation. _____

    AFTER RELEASE FROM CLAMPS, ETC

31. Explain the meaning of the free state symbol placed in the feature control frame after the geometric tolerance and material condition symbol (if any material condition symbol is used). _____

32. Describe a composite profile tolerance. _____

    CONTROL FORM & ORIENTATION

Print Reading Exercise

Name: _____

5

The following print reading exercise uses actual industry prints with related questions that require you to read specific dimensioning and geometric tolerancing representations. The answers should be based on previously learned content of this book. The prints used are based on ASME standards, however company standards may differ slightly. When reading these prints, or any other industry prints, a degree of flexibility may be required to determine how individual applications correlate with the ASME standards.

Print Reading Exercise

Refer to the print of the PLATE-TOP MOUNTING found on page 311.

1. Name at least one form and one profile geometric tolerances found on this print.

    PROFILE: ⌒ .010 A   FORM: ⌰ .005

    B   A

2. Completely describe the form geometric tolerances found on this print. _____

    FLATNESS @ A w/in .005

    PERPENDICULARITY @ C w/in ⌀.002 TO B

3. Completely describe the profile geometric tolerances found on this print. _____

    PROFILE OF SURFACE @ B w/in .010 TO A

Refer to the print of the HOUSING-LENS, FOCUS found on page 313.

4. Name the profile geometric tolerances found on this print. PROFILE OF SURFACE & LINE

5. Completely describe the profile geometric tolerances and related features found on this print.

    a) SURF 2x R.930   ⌒ .005 A B C

    LINE   R

b)   2X .120   ⌒ | .002 | A | B | C

c)   2 X R 1.060   ⌒ | .002 | A | B | C

6. What does the title block specify in regard to form geometric tolerances? _____

Chapter 6
Tolerances of Orientation and Runout

This chapter explains the concepts and techniques of dimensioning and tolerancing to control the orientation and runout of geometric shapes.

Orientation geometric tolerances control:

❏ Parallelism.
❏ Perpendicularity.
❏ Angularity.

Runout is a combination of controls that may include any of the following:

❏ The control of circular elements of a surface.
❏ Control of the cumulative variations of circularity, straightness, coaxiality, angularity, taper, and profile of a surface.
❏ Control of variations in perpendicularity and flatness.

When size tolerances provided in conventional dimensioning do not provide sufficient control for the functional design and interchangeability of a product, then form and/or profile tolerances should be specified. Size limits control a degree of form and parallelism. Locational tolerances control a certain amount of orientation. Therefore, the need for further form and orientation control should be evaluated before specifying geometric tolerances of form and orientation.

ORIENTATION TOLERANCES

Orientation tolerances control the relationship of features to one another. *Orientation tolerances* include parallelism, perpendicularity, angularity, and profile (in some cases). When controlling orientation tolerances, the feature is related to one or more datum features. Relation to more than one datum should be considered if required to stabilize the tolerance zone in more than one direction. Parallelism, perpendicularity, and angularity tolerances control flatness in addition to their intended orientation control.

Orientation tolerances are total. This means that all elements of the related surface or axis fall within the specified tolerance zone. When less demanding requirements controlling only individual line elements of a surface meet the design goal, then a note such as "EACH ELEMENT" or "EACH RADIAL ELEMENT" should be shown below the associated feature control frame. This application permits individual line elements of a surface, rather than the total surface, to be controlled in relation to a datum.

Orientation tolerances imply RFS. Therefore, MMC or LMC must be specified if any application other than RFS is intended.

Parallelism Tolerance

A *parallelism geometric tolerance* is two parallel planes or cylindrical zones that are parallel to a datum plane where the surface or axis of the feature must lie. The parallelism geometric characteristic symbol and associated feature control frame is detailed in Example 6-1.

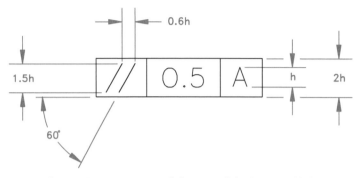

Example 6-1. Feature control frame with the parallelism geometric characteristic symbol and datum reference.

Surface parallelism

When a surface is to be parallel to a datum, the feature control frame is either connected by a leader to the surface, or connected to an extension line from the surface. The actual surface must be within the parallelism tolerance zone that is two planes parallel to the datum. The parallelism tolerance zone must be within the specified size limits. Refer to Example 6-2.

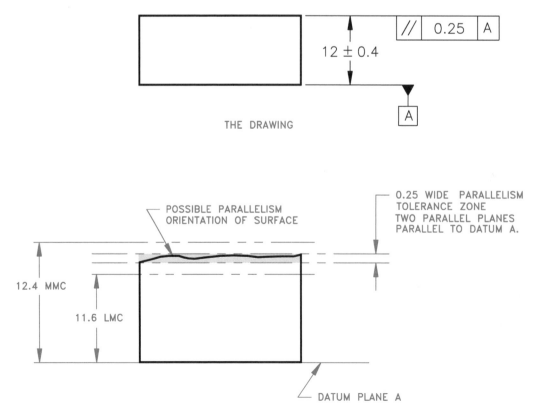

Example 6-2. Application of the parallelism geometric tolerance.

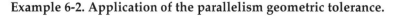

Axis parallelism

A parallelism tolerance may be two parallel planes that are parallel to a datum plane where the axis of a feature must lie. For example, the axis of a hole may be specified within a tolerance zone that is parallel to a given datum. This parallelism tolerance zone must also be within the specified locational tolerance. The feature control frame is placed with the diameter dimension, as shown in Example 6-3. Remember, placing the feature control frame with a diameter dimension associates the related geometric tolerance with the feature axis. RFS is assumed.

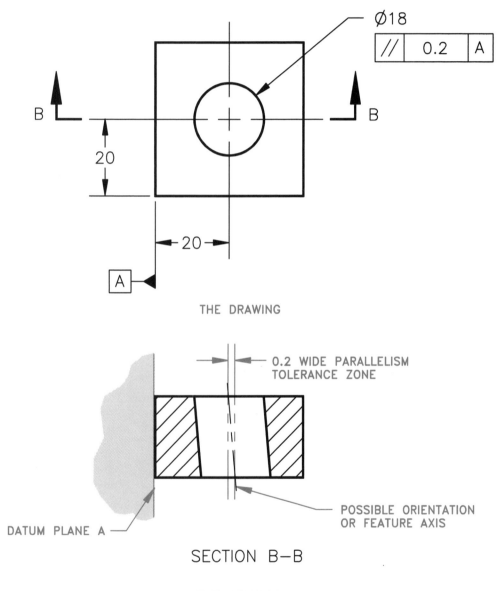

THE DRAWING

SECTION B–B

THE MEANING

Example 6-3. Specifying parallelism of an axis to a datum plane.

Parallelism may also be applied to the axis of two or more features when a parallel relationship between the features is desired. The axis of a feature must lie within a cylindrical tolerance zone that is parallel to a given datum axis. This is a diameter tolerance zone, as shown in Example 6-4.

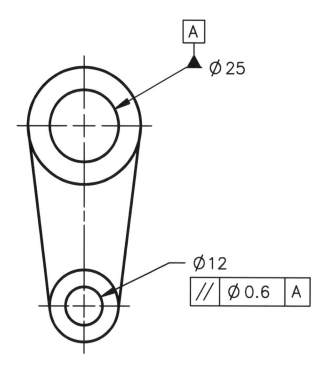

THE DRAWING

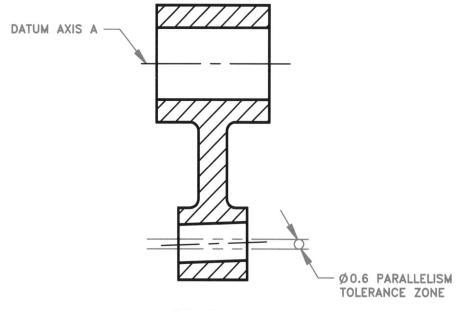

THE MEANING

Example 6-4. Specifying parallelism of an axis to a datum axis.

Parallelism of line elements

Orientation tolerances (parallelism, perpendicularity, angularity, and, in some cases, profile) are implied to be total where an axis or all elements of a surface must fall within the specified tolerance zone. Where it is desirable to control only individual line elements, rather than the entire surface, the note "EACH ELEMENT" or "EACH RADIAL ELEMENT" is placed below the feature control frame, as shown in Example 6-5.

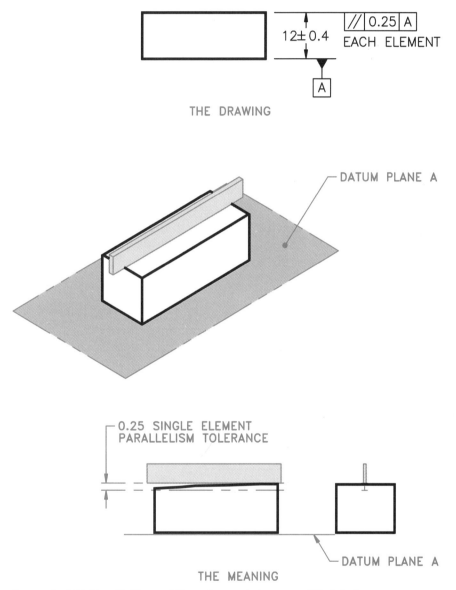

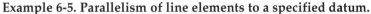

Example 6-5. Parallelism of line elements to a specified datum.

Perpendicularity Tolerance

A *perpendicularity tolerance* is established by a specified geometric tolerance zone made up of two parallel planes or cylindrical zones that are a basic 90° to a given datum plane or axis where the actual surface or axis must lie. The perpendicularity geometric characteristic symbol placed in a feature control frame is detailed in Example 6-6.

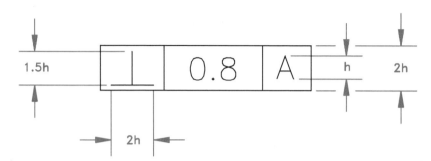

Example 6-6. Feature control frame with the perpendicularity geometric characteristic symbol and datum reference.

Perpendicularity of a surface

The feature control frame may be connected to the surface with a leader or from an extension line, as shown in Example 6-7. The actual surface is oriented between two parallel planes that are perfectly 90° to the datum plane.

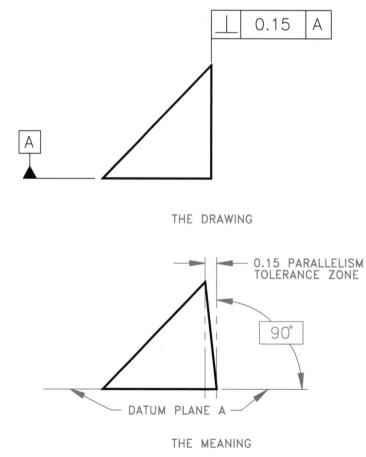

Example 6-7. Application of the perpendicularity geometric tolerance to one datum reference.

In Example 6-7, the surface is held perpendicular to one datum plane. It is also possible to hold a surface perpendicular to two datum planes. When this is done, the surface must lie between two parallel planes that are perpendicular to two datum planes. Both datums are referenced in the feature control frame, as shown in Example 6-8.

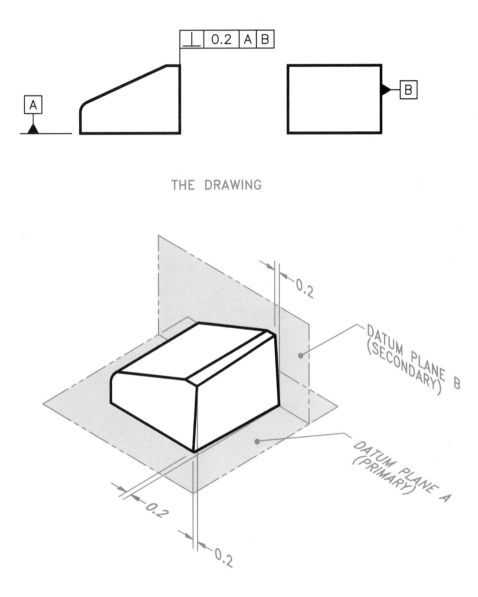

THE DRAWING

THE MEANING

Example 6-8. Application of the perpendicularity geometric tolerance to two datum references.

Perpendicularity of an axis

Perpendicularity may be a tolerance zone made up of two parallel planes perpendicular to a datum plane or axis where the axis of a feature must lie. For example, the axis of a hole may be between two parallel planes that are perpendicular to a datum axis. In this application, the datum axis is established and the feature control frame is placed below the diameter dimension controlling perpendicularity, as shown in Example 6-9. RFS is implied unless MMC or LMC is placed in the feature control frame after the geometric tolerance.

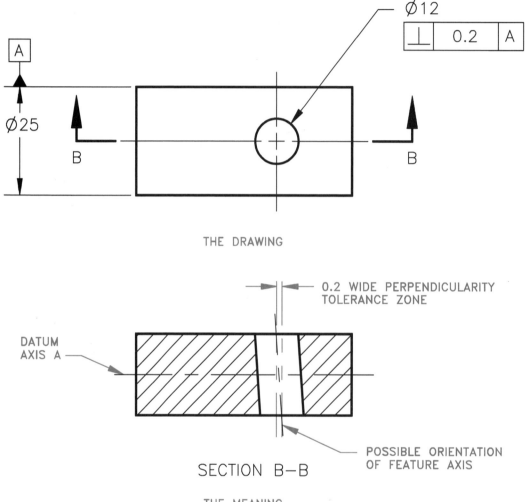

Example 6-9. Specifying perpendicularity of an axis to a datum axis.

Another application that may require a perpendicularity specification is a cylindrical feature such as a pin or stud. In this situation, the feature axis is within a cylindrical tolerance zone that is perpendicular to a datum plane. The feature control frame is attached to the diameter dimension and a diameter symbol precedes the geometric tolerance to specify a cylindrical tolerance zone, as shown in Example 6-10.

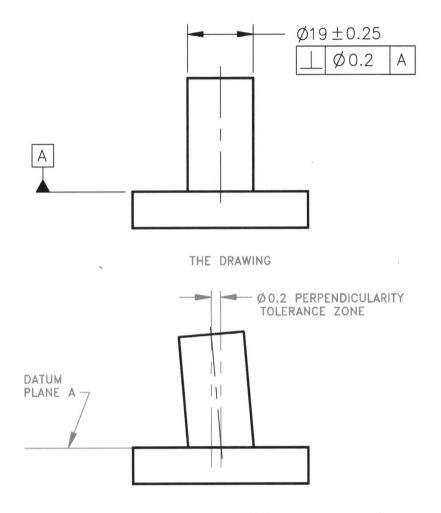

Example 6-10. Specifying the perpendicularity of an axis to a datum plane.

Perpendicularity of a center plane

A symmetrical feature, such as a slot, may be specified as perpendicular to a datum plane. In this application, the feature center plane is held within two parallel planes that are perpendicular to a given datum plane. The center plane must also be within the specified locational tolerance. Refer to Example 6-11.

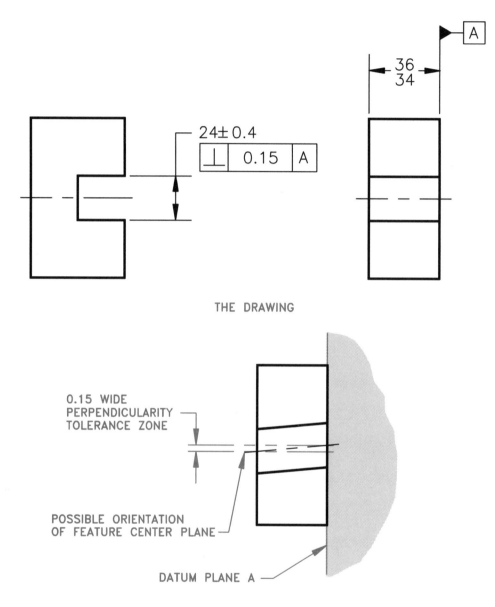

Example 6-11. Specifying the perpendicularity of a center plane to a datum plane.

Perpendicularity of line elements

Another possibility is that single line elements of a surface, rather than the entire surface, may be perpendicular to a given datum. When any single line element of the object shall be held perpendicular to a datum, the words "EACH ELEMENT" are indicated below the feature control frame, as shown in Example 6-12.

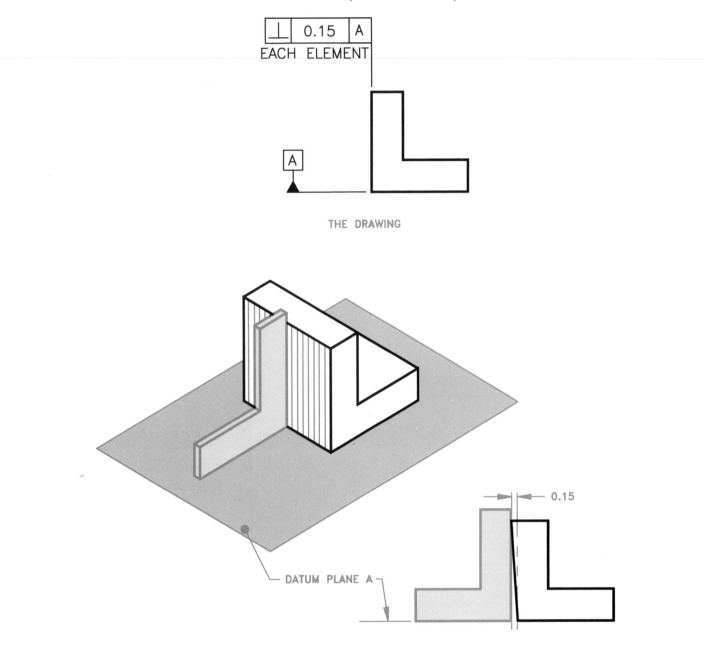

THE DRAWING

THE MEANING

Example 6-12. Perpendicularity of line elements to a datum plane.

Angularity Tolerance

An *angularity geometric tolerance* zone is established by two parallel planes or cylindrical zones at any specified basic angle, other than 90°, to a datum plane, a pair of datum planes, or an axis. The angularity geometric characteristic symbol placed in a feature control frame is detailed in Example 6-13.

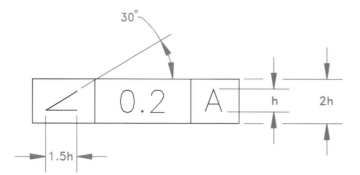

Example 6-13. Feature control frame with the angularity geometric characteristic symbol and datum reference.

Angularity of a surface

The feature control frame is normally connected to the surface by a leader. The specified angle must be basic and must be dimensioned from the datum plane, as shown in Example 6-14. RFS is implied unless otherwise specified.

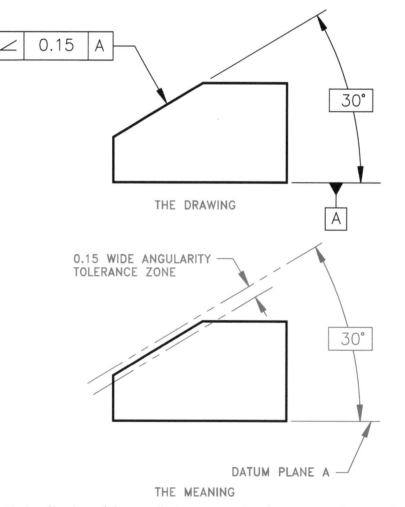

Example 6-14. Application of the angularity geometric tolerance to a datum reference.

Angularity of an axis

The axis of a hole or other cylindrical feature can be dimensioned with an angularity tolerance if the feature is at an angle other than 90° to a datum plane or axis. This specification establishes two parallel planes spaced equally on each side of the specified basic angle from a datum plane or axis where the axis of the considered feature must lie. This control applies only to the view where it is specified. The feature control frame is shown next to the feature diameter dimension to specify axis control, as shown in Example 6-15.

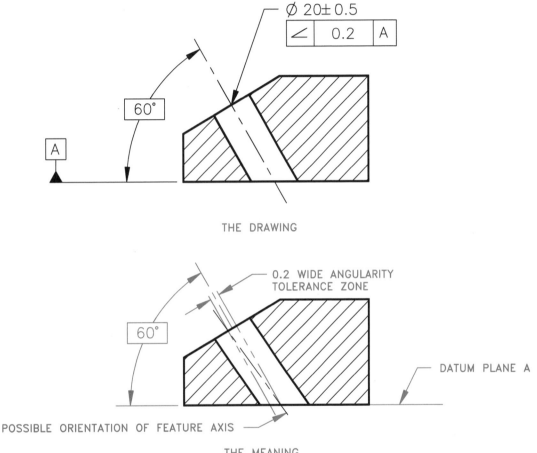

Example 6-15. Application of the angularity of an axis to a datum plane.

It is also possible to control the feature axis within a cylindrical angularity tolerance zone. To do this, a diameter symbol is placed in front of the geometric tolerance in the feature control frame. This indicates that the angularity tolerance zone is cylindrical, as shown in Example 6-16.

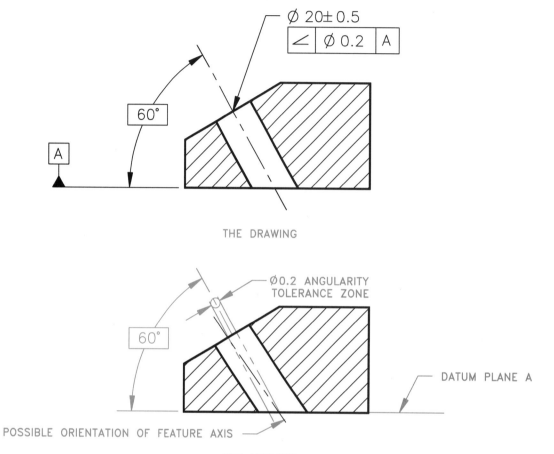

Example 6-16. Controlling the axis of a feature within a cylindrical angularity tolerance to a datum reference.

Angularity of a center plane and single element control

An angularity geometric tolerance, just like parallelism and perpendicularity, may control the orientation of the center plane of a symmetrical feature, such as a slot or center plane. This angularity tolerance is formed by two parallel planes at a specified basic angle to a datum plane where the center plane of the feature must lie.

Angularity may also be controlled on a single line element or single radial element basis by placing the words "EACH ELEMENT" or "EACH RADIAL ELEMENT" below the feature control frame.

Application of Orientation Tolerances at RFS, MMC, and Zero Tolerance at MMC

When a material condition symbol is *not* shown in the feature control frame, then RFS is implied. This means that the geometric tolerance is the same when the feature is manufactured at any produced size.

Placing the MMC material condition symbol after the geometric tolerance in the feature control frame means that the tolerance is held at the MMC produced size, and then the geometric tolerance is allowed to increase equal to the amount of departure from MMC.

Another application is zero geometric tolerance at MMC. This is done when the geometric tolerance in the feature control frame is zero and the MMC material condition symbol is used. This means that at the MMC produced size, the feature must be perfect in orientation with respect to the specified datum. As the actual produced size departs from MMC, the geometric tolerance increases equal to the amount of departure.

This is done to create a boundary of perfect form at MMC to control the relationship between features. Zero orientation tolerance at MMC may be used for parallelism, perpendicularity, or angularity. While the zero geometric tolerance at MMC method is easy and eliminates the use of a note, the note "PERFECT ORIENTATION AT MMC REQUIRED FOR RELATED FEATURES" may also appear on the drawing.

Applications of RFS, MMC, and zero tolerance at MMC are shown in Example 6-17.

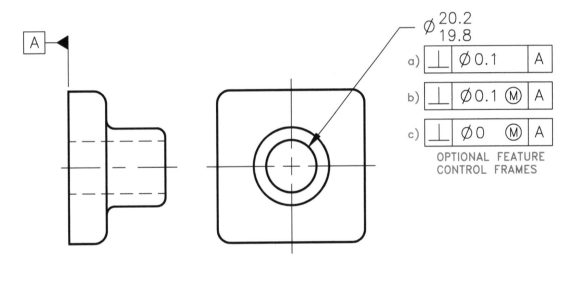

GEOMETRIC TOLERANCES BASED ON THE OPTIONAL FEATURE CONTROL FRAMES

	POSSIBLE PRODUCED SIZE	a) RFS	b) MMC	c) ZERO AT MMC
MMC	19.8	0.1	0.1	0
	19.9	0.1	0.2	0.1
	20.0	0.1	0.3	0.2
	20.1	0.1	0.4	0.3
LMC	20.2	0.1	0.5	0.4

Example 6-17. Application of orientation geometric tolerances at RFS, MMC, and zero orientation tolerances at MMC.

RUNOUT TOLERANCE

Runout is a combination of geometric tolerances used to control the relationship of one or more features of a part to a datum axis. Features that may be controlled by runout are either surfaces around, or perpendicular to, a datum axis. The datum axis should be selected as a diameter of sufficient length, as two diameters adequately separated on the same axis, or as a diameter and perpendicular surface. There are two types of runout–total runout and circular runout. The type of runout selected depends on design and manufacturing considerations. Circular runout is generally a less complex requirement than total runout. The feature control frame is connected by a leader line to the surface. Multiple leaders may be used to direct a feature control frame to two or more surfaces having a common runout tolerance. The runout geometric characteristic symbols are shown detailed in feature control frames in Example 6-18.

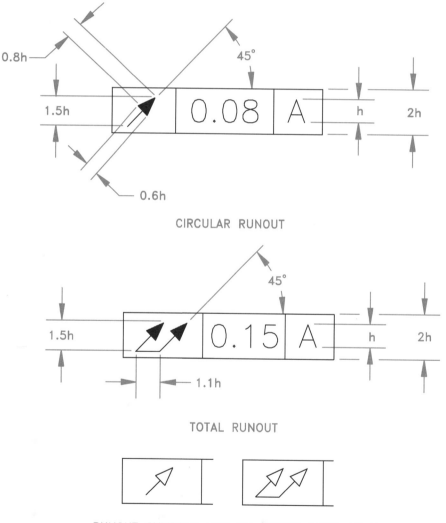

Example 6-18. Feature control frames with the circular and total runout geometric characteristic symbols and datum references. The runout symbol arrows may be filled or unfilled, depending on company preference.

Circular Runout

Circular runout provides control of single circular elements of a surface. When applied to surfaces around a datum axis, circular runout controls circularity and coaxiality. *Coaxiality* is a situation where two or more features share a common axis. When applied to surfaces at right angles to a datum axis, circular runout may be used to control wobbling motion. This tolerance is measured by the full indicator movement (FIM) of a dial indicator placed at several circular measuring positions as the part is rotated 360°. FIM shows a total tolerance. Refer to Example 6-19. An example of circular runout is shown in Example 6-20.

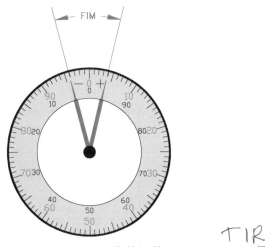

TIR

Example 6-19. A dial indicator showing full indicator movement (FIM).

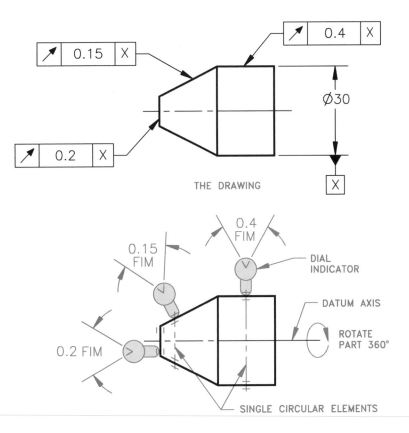

Example 6-20. The application of circular runout.

Total Runout

SPIN ON AXIS, SO MORE ACCURATELY DEFINES PART SHAPE

Total runout provides a combined control of a surface element. This is a tolerance that blankets the surface to be controlled. Total runout is used to control the combined variations of circularity, straightness, coaxiality, angularity, taper, and profile when applied to surfaces around a datum axis. Total runout may be used to control the combined variations of perpendicularity (to control wobble and flatness) and to control concavity or convexity when applied to surfaces perpendicular to a datum axis. The total runout tolerance zone encompasses the entire surface as the part is rotated 360°. The entire surface must lie within the specified tolerance zone. In order to determine this, the dial indicator is placed at selected locations along the surface as the part is rotated 360°. Total runout is shown in Example 6-21.

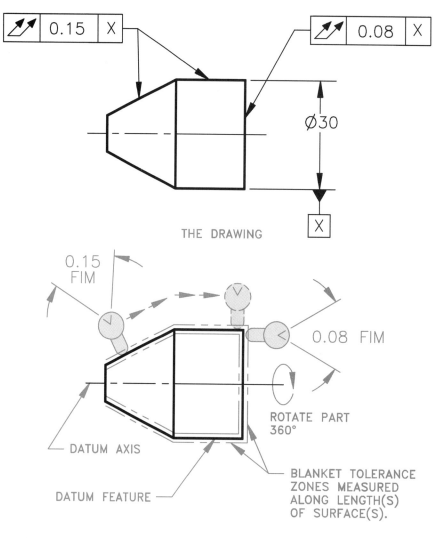

Example 6-21. The application of total runout.

Runout to a Portion of a Surface and Two Datum References

A portion of a surface may have a runout tolerance specified if it is not desired to control the entire surface. This is done by placing a chain line located with basic dimensions in the linear view. The *chain line* identifies where the profile tolerance around the object is located. The feature control frame is then connected by a leader to the chain line. Runout tolerances may also be applied where two datum diameters collectively establish a single datum axis. This is done by placing the datum identifying letters separated by a dash in the feature control frame. For example, "G-H" shown in Example 6-22 is how this would appear on a drawing.

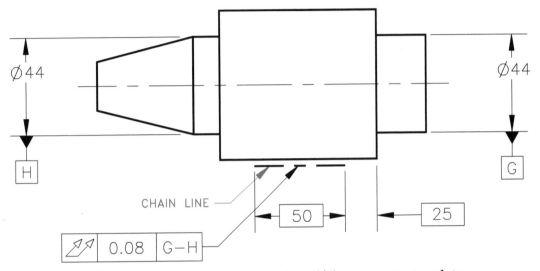

Example 6-22. Using partial surface runout and specifying runout to two datum references together.

Runout to a Datum Surface and a Datum Axis

Runout geometric tolerances may be controlled in relationship to a datum axis and a surface at right angles to the axis. When this is done the datums are placed separately in the feature control frame in their order of precedence. Each circular element (for circular runout) or each surface (for total runout) must be within the specified geometric tolerance when the part is mounted on the datum surface and rotated 360° about the datum axis. Refer to Example 6-23.

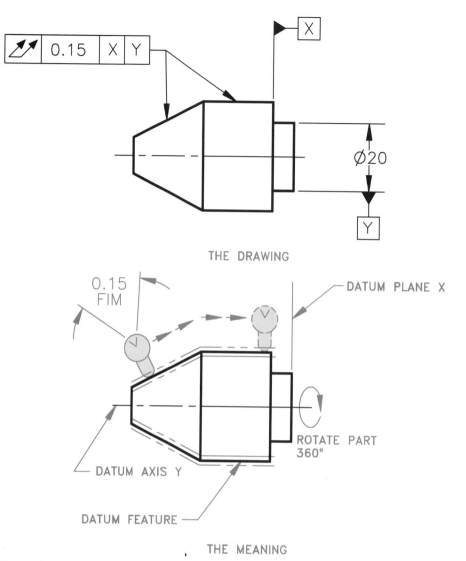

Example 6-23. Specifying runout to a datum surface and a datum axis.

Runout Control to a Datum

When a datum feature symbol is specified on a runout control, then the runout geometric tolerance applies to the datum feature. This is done by centering the datum feature symbol below the feature control frame or connecting the datum feature symbol to the leader shoulder, as shown in Example 6-24.

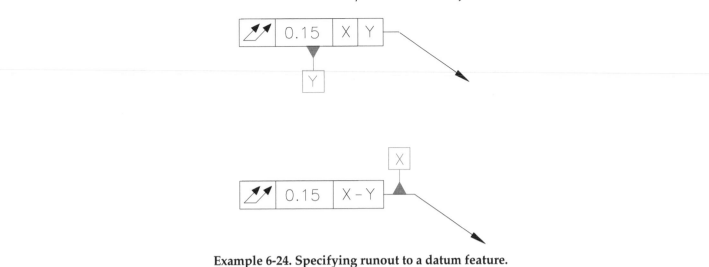

Example 6-24. Specifying runout to a datum feature.

COMBINATION OF GEOMETRIC TOLERANCES

Profile tolerancing may be combined with other types of geometric tolerances. For example, a surface may have a profile tolerance controlled within a specified amount of parallelism relative to a datum. When this is done, the surface must be within the profile tolerance, and each line element of the surface must be parallel to the given datum by the specified parallelism tolerance, as shown in Example 6-25.

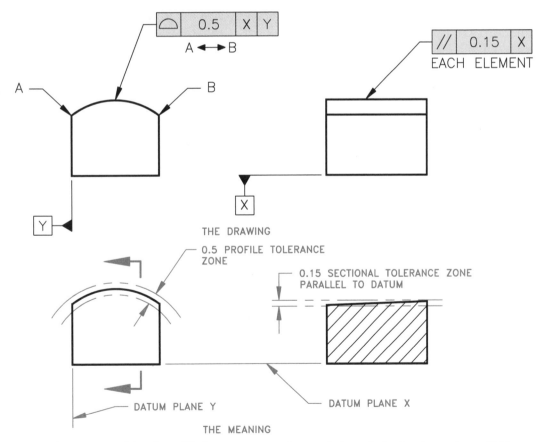

Example 6-25. Combined profile between two given points and single element parallelism geometric tolerance.

A surface may also be controlled by profile and refined by runout. When this is done, as shown in Example 6-26, any line element of the surface must be within the profile tolerance, and any circular element of the surface must be within the specified runout tolerance.

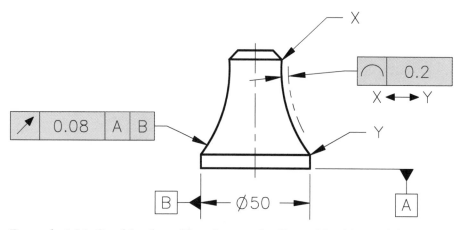

Example 6-26. Combined profile tolerance (unilateral in this case) between two points and circular runout.

In some situations, it may be necessary to control runout constrained by flatness, straightness, or cylindricity. A combination runout and cylindricity, as shown in Example 6-27, means that the datum surface must be controlled within the specified tolerances of runout and cylindricity. Notice that the different feature control frames are attached. This is different from the previous examples where the feature control frames were separate.

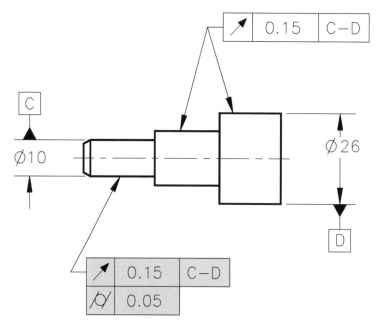

Example 6-27. Combined runout to two coaxial datums and cylindricity.

The combination of perpendicularity and parallelism may be achieved by combining the perpendicularity and parallelism controls, as shown in Example 6-28. This allows versatility by providing uniform perpendicularity and parallelism to related datums.

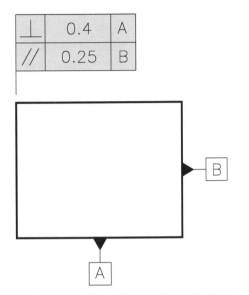

Example 6-28. Combined perpendicularity and parallelism to the same surface.

The tolerance zones are often different, but when appropriate they may be the same. When the geometric tolerance zones are the same, they still remain in their own feature control frame compartments, as shown in Example 6-29.

⊥	0.25	A
//	0.25	B

Example 6-29. Combined feature control frame with equal geometric tolerance zones.

USING THE TANGENT PLANE SYMBOL

Geometric tolerance zones are total. This means that the axis or center plane, or all surface elements must fall within this zone, unless the note "EACH ELEMENT" or "EACH RADIAL ELEMENT" is placed below the feature control frame as discussed earlier in this chapter. The total geometric tolerance zone also means that the axis or center plane, or surface elements may be anywhere within the geometric tolerance zone. An additional requirement may be applied to the surface within the geometric tolerance zone by placing the tangent plane symbol after the geometric tolerance in the feature control frame. A *tangent plane* is a theoretically exact plane that is established by the true geometric counterpart of the feature surface. Refer to Example 6-30. When the tangent plane symbol is used, this means that a plane contacting the high points of the surface must be within the specified geometric tolerance zone, as shown in Example 6-30. It is possible for elements of the surface to fall outside the geometric tolerance zone with this specification.

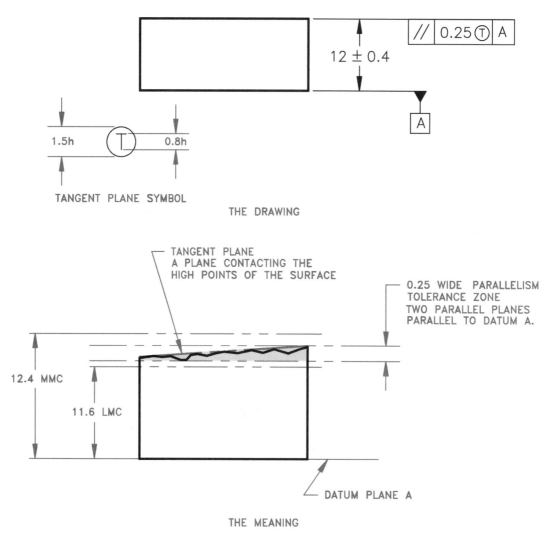

Example 6-30. Using the tangent plane application.

Name: _____

1. Orientation geometric tolerances control ___PARALLEL___,
   ___PERP.___, and ___ANGULARITY___.

2. Runout is a combination of controls that may include:

   a)   control of circular elements of a surface.

   b)   control of the cumulative variations of circularity, straightness, coaxiality, angularity, taper, and profile of a surface.

   c)   control variations of perpendicularity and flatness.

   (d)) all of the above.

3. ___ORIENTATION___ tolerances control the relationship of features to one another.

4. Orientation tolerances must be related to one or more datum features. (True) or False.

5. Orientation tolerances ⌐COULD control flatness. True or False.

6. Which of the following does an orientation tolerance imply: MMC, RFS, or LMC? ___RFS   ALWAYS___

7. A(n) ___PARALLELISM___ tolerance zone is the distance between two parallel planes parallel to a datum.

8. What does the placement of a parallelism feature control frame below a diameter dimension mean? ___AXIS CONTROL___
   _____
   _____
   _____
   _____

9. Parallelism may be applied to the axis of two or more features when a parallel relationship between the features is desired. (True) or False?

10. Orientation tolerances are implied to be total. Therefore, how must a drawing be modified where it is desirable to control only individual line elements rather than the entire surface? _____
    ___EACH RADIAL ELEMENT___
    _____
    _____
    _____

11. ___⊥___ is established by a geometric tolerance zone made up of two parallel planes that are a basic 90° to a given datum plane or axis where the actual surface must lie.

12. A symmetrical feature, such as a slot, may be specified as perpendicular to a datum plane. In this application, the feature CENTER PLANE is held within two parallel planes that are ___⊥___ to a given datum.

13. A(n) __ANGULARITY__ tolerance zone is established by two parallel planes at any specified basic angle, other than 90°, to a datum plane or axis. The specified angle must be ___BASIC___ and must be dimensioned from the ___DATUM___ plane.

14. Given the following drawing, a reference chart showing a range of possible produced sizes, and three (3) optional feature control frames that may be applied to the diameter dimension, provide the geometric tolerance at each possible produced size for each feature control frame application.
    Suggestion: Review Chapter 4 *Material Condition Symbols*.

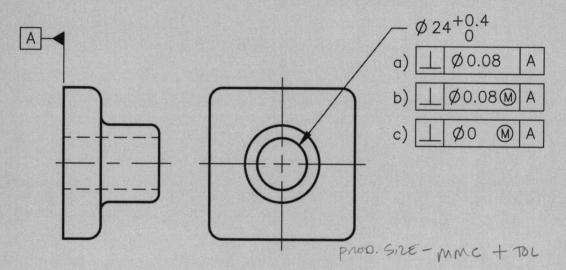

$\emptyset 24^{+0.4}_{\ 0}$

a) | ⊥ | Ø 0.08 | A |

b) | ⊥ | Ø 0.08 Ⓜ | A |

c) | ⊥ | Ø 0 Ⓜ | A |

PROD. SIZE — MMC + TOL

DIAMETER TOLERANCE ZONES ALLOWED

Possible Produced Sizes	a) RFS	b) MMC	c) Zero at MMC
24.0	.08	.08	0
24.1	.08	.18	0.1
24.2	.08	.28	.2
24.3	.08	.38	.3
24.4	.08	.48	.4

15. __RUNOUT__ is a combination of geometric tolerances used to control the relationship of one or more features of a part to a datum axis.

16. Features that may be controlled by runout are either surfaces around or perpendicular to a datum axis. (True) or False.

17. The two types of runout are ___CIRCULAR___ and ___TOTAL___.

18. __CIRCULAR__ provides control of single circular elements of a surface.

19. When applied to surfaces around a datum axis, circular runout controls ___CIRCULARITY___ and ___CO-AXIALITY___.

20. _COAXIALITY_ is a situation where two or more features share a common axis.

21. _TOTAL RUNOUT_ provides a combined control of surface elements.

22. ___"___ is used to control the combined variations of circularity, straightness, coaxiality, angularity, taper, and profile when applied to surfaces around a(n) _DATUM_ axis.

23. _TOTAL RUN-OUT_ may be used to control the combined variations of perpendicularity, to control wobble and flatness, or to control concavity or convexity when applied to surfaces perpendicular to a datum axis.

24. Explain the fundamental difference between how circular and total runout are established. _CIRCULAR – SINGLE CIRCULAR LOCATION_

25. What does the chain line mean when specifying runout to a portion of a surface? _NOTE LOCATION NEC. FOR TOL._

26. Explain the combination of geometric tolerances that exist in the following drawing. _BETWEEN A & B PROFILE W/IN .5_
// .15 w/ RESPECT TO X

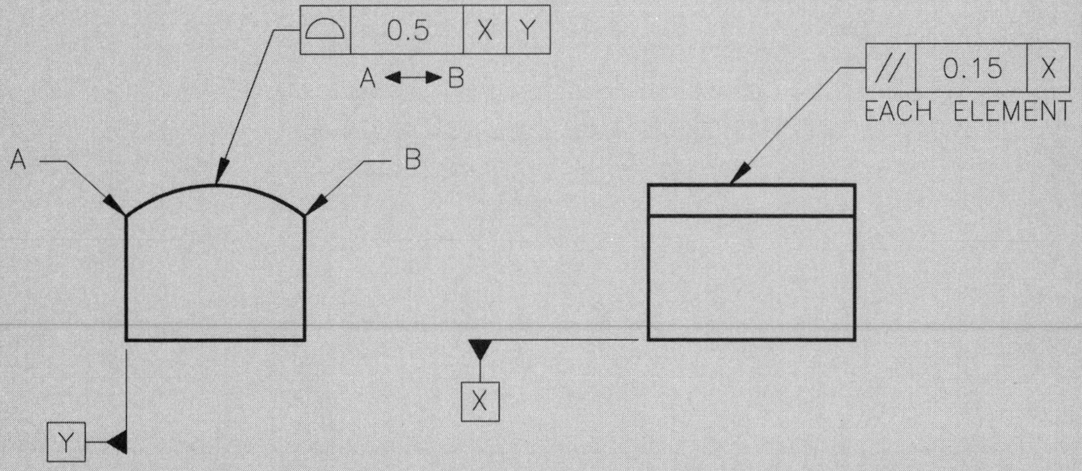

| ⌒ | 0.5 | X | Y |

A ⟷ B

| // | 0.15 | X |

EACH ELEMENT

A

B

X

Y

27. Explain the combination of geometric tolerances that exist in the following drawing.

_____

_____

_____

_____

_____

_____

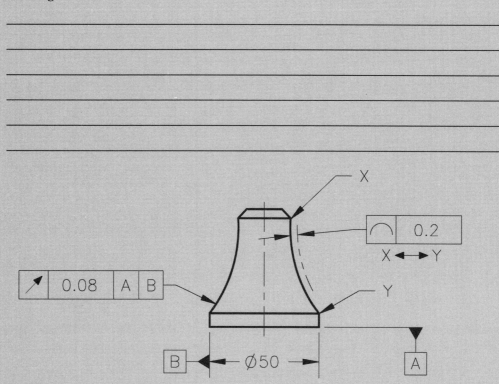

28. Given the drawing below, explain the meaning represented by the following specifications:

a) Datum feature symbol C.   Ø OF 10   DATUM AXIS

b) Datum feature symbol D.   " " 26 " "

c) Datum reference C-D in the feature control frame with the runout geometric tolerance.

d) Combination runout and cylindricity.

MUST BE CONTROLLED
W/IN .15 RUNOUT
     .05 CYLINDRICITY

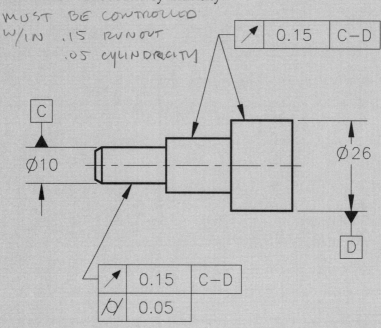

a) _____

b) _____

c) _____

   _____

d) _____

   _____

29. Name the geometric tolerances that may be used to control orientation: _____

   _____

   _____

   _____

30. Explain the geometric tolerance that exists in the following drawing.

   _____

   _____ // A w/in 0,2 _____

   _____

   _____

   _____

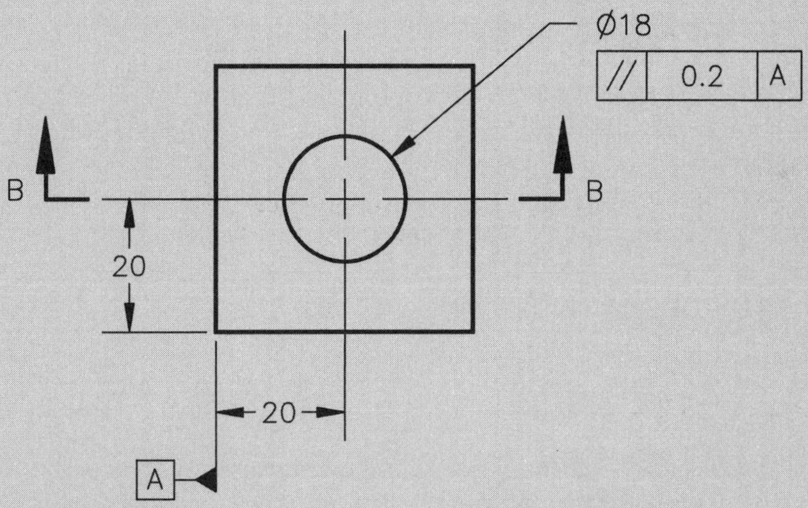

   _____

Print Reading Exercise

Name: _____

6

The following print reading exercise uses actual industry prints with related questions that require you to read specific dimensioning and geometric tolerancing representations. The answers should be based on previously learned content of this book. The prints used are based on ASME standards, however company standards may differ slightly. When reading these prints, or any other industry prints, a degree of flexibility may be required to determine how individual applications correlate with the ASME standards.

Print Reading Exercise

Refer to the print of the SLEEVE-DEWAR REIMAGING found on page 303.

1. Name at least three orientation and runout geometric tolerances found on this print. ___↗↗___ @ 1.274∅___
   ___⊥___ .874 ∅___
   ___∥ @ 2.059___
   _____

2. Completely describe the orientation geometric tolerances and related features found on this print, and include if the geometric tolerance is a surface, axis, or center plane control, and give the material condition symbol applied to the geometric tolerance. ___
   ___2.059" FROM -A- & ∥ TO A W/IN .001 RFS___
   _____
   AXIS OF ___∅.8740 ⊥ to -A- W/IN .001 RFS___
   _____
   SURFACE OF 1.274_____

3. Completely describe the runout geometric tolerances and related features found on this print, and include if the geometric tolerance is a surface, axis, or center plane control, and give the material condition symbol applied to the geometric tolerance. ___
   ___↗↗   SURFACE   @ 1.274" ∅___
   _____
   _____
   _____
   _____

4. What is the maximum FIM allowed in regard to the runout geometric tolerance identified in question number 3? _____
   ___.010___

5. Briefly describe how the geometric tolerance specified in question number 3 is inspected. _____

   HOLD PART BTWN CTRS & ROTATE 360°

Refer to the print of the HUB-STATIONARY ATU found on page 305.

6. Name at least three different orientation and runout geometric tolerances found on this print. ⟂ , ∠ , ∥ .

7. Completely describe the <u>orientation geometric tolerances</u> and related features found on this print, and include if the geometric tolerance is a surface, axis, or center plane control, and give the material condition symbol applied to the geometric tolerance.

   ⌀ 1.250 ± .002 to BE ∥ W/IN .002 OF A

8. Refer to the ⌀4.4997/4.4994 feature at Datum C.

   a) Completely describe the geometric tolerances associated with this dimension.

   ⟂

   b) What does the reference to Datum B-C mean? _____

   CO-AXIAL DATUM AXIS

   c) Is the datum reference identified in question 8b) primary, secondary, or tertiary?

   SECONDARY

9. What is the maximum FIM allowed in regard to the runout geometric tolerance identified in question number 8? .0004

10. Briefly describe how the geometric tolerance specified in question number 8 is inspected. _____

_____ PANT BTWN CTRS OF AXIS B & ROTATE 360°_____

_____ & NOT EXCEED .0004 FIM (TIR)_____

_____

_____

11. Refer to the R1.525 dimension.

    a) How many features relate to this dimension? _____ 6   TOTAL PLC S

    b) Completely describe the geometric tolerance associated with this dimension.

    _____

    _____

    _____

    _____

Refer to the print of the PLATE-TOP MOUNTING found on page 311.

12. Name at least three different orientation and runout geometric tolerances found on this print. _____ ⟋ , ⊥ , ∠ _____

    _____

    _____

    _____

13. Completely describe the orientation geometric tolerances and related features found on this print, and include if the geometric tolerance is a surface, axis, or center plane control, and give the material condition symbol applied to the geometric tolerance.

    a) _____ 12 PLCS ∠   (INCLUDING 0° @ LH SIDE)_____

    _____ FROM DATUM AXIS D_____

    _____

    b) _____ AXIS OF 7.400 ⊥ B w/ CYLD, TOL_____

    _____ ZONE .002 @ MMC_____

    _____

    _____

14. Refer to the Ø6.750 feature. Completely describe the geometric tolerances associated with this dimension. _____

    _____ ⟋ FOR SURFACE OF FEATURE OF .010_____

    _____ w/ RESPECT TO -C- AXIS_____

    _____

    _____

Refer to the print of the B.H.-TOP found on page 312.

15. Refer to the Ø24.13/24.10 dimension.

   a) Completely describe the geometric tolerance associated with this feature.

   ⊥ .5 RFS w/ RESPECT TO A & B

   "⊥" AROUND

   b) Give the geometric tolerance applied to the following list of possible produced sizes:

Produced Sizes	Geometric Tolerances	
24.13	0.5	*BECAUSE RFS*
24.12	0.5	
24.11	0.5	
24.10	0.5	

16. Refer to the tapered feature with the dimensions Ø23.34 and 8°26′:

   a) Completely describe the geometric tolerance associated with this feature.

   b) Explain the function of the datum references associated with this geometric tolerance. *⊥ to A & AROUND B*

Refer to the print of the HOUSING-LENS, FOCUS found on page 313.

17. Completely describe the specifications provided in the feature control frame associated with Datum B. *SEE R.H. VIEW*

   ⊥ to A w/in .002

18. What does the title block refer to in regard to orientation?

Refer to the print of the PLATE-BOTTOM WEDGED, ADJUSTABLE PARALLEL (HP) found on page 314.

19. Refer to the angularity geometric tolerance.

    a) What is the geometric tolerance? ___.002___

    b) What is the material condition? ___RFS___

    c) Name the reference datum. ___A___

    d) Give the angle from the reference datum. ___BASIC 10°___

    e) Describe the combination geometric tolerance associated with this feature.

    ___∠ .001   RFS___

    _____

    _____

    _____

    _____

20. Refer to the 2.10 dimension.

    a) Name the line type that is made up of alternating long and two short dashes that this dimension controls. ___CHAIN LINE___

    b) What is the purpose of this type of presentation? _____
    ___SPECIFIC   CONTROL   OVER   THIS   AREA___

    _____

    _____

21. Describe the combination orientation geometric tolerance applied to Datum C.

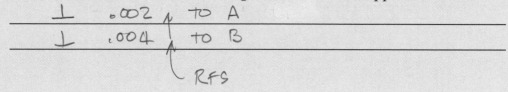

    ___⊥   .002 ↑ TO A___
    ___⊥   .004 ↑ to B___
    ___RFS___

Chapter 7
Location Tolerances (Part I)

Location tolerances are used for the purpose of locating features from datums or for establishing coaxiality or symmetry. Location tolerances include:

❑ Positional.
❑ Concentricity.
❑ Symmetry.

Positional tolerancing is used to define a zone where the center, axis, or center plane of a feature of size is permitted to vary from true position. *True position* is the theoretically exact location of a feature. *Basic dimensions* are used with datum or chain dimensioning systems to establish the true position from specified datum features and between interrelated features.

Location tolerancing is specified by a position, a concentricity or symmetry symbol, a tolerance, and appropriate datum references placed in a feature control frame. When positional tolerancing is used, the MMC or LMC material condition symbols must be specified after the tolerance and applicable datum reference in the feature control frame. Otherwise, RFS is assumed.

In comparison to conventional methods, the use of positional tolerancing concepts provides some of the greatest advantages to mass production. The coordinate dimensioning system limits the actual location of features to a rectangular tolerance zone. Using positional tolerancing, the location tolerance zone changes to a cylindrical shape, thus increasing the possible location of the feature by about 57%. This improves the interchangeability of parts while increasing manufacturing flexibility and reducing the scrappage of parts. The use of MMC applied to the positional tolerance allows the tolerance zone to increase in diameter as the feature size departs from MMC. This application also allows greater flexibility in the acceptance of mating parts.

POSITIONAL TOLERANCE

The positional geometric characteristic symbol is placed in the feature control frame as shown in Example 7-1. The next compartment of the feature control frame contains the diameter symbol (if a cylindrical tolerance zone is applied) followed by the specified positional tolerance and a material condition symbol (if MMC or LMC is used). Additional compartments are used for datum reference.

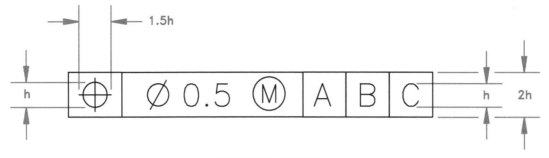

h = LETTERING HEIGHT

Example 7-1. Positional geometric characteristic symbol and tolerance in a feature control frame.

Some specific characteristics of positional tolerances include the following.

❏ *True position* is the theoretically exact location of the centerline of a feature, as shown in Example 7-2.

❏ A material condition symbol at MMC or LMC must follow the specified positional tolerance, datum reference, or both as needed. Otherwise, RFS is assumed.

❏ Positional tolerances control the location of a cylindrical tolerance zone where the centerline of a feature is located, as shown in Example 7-2. Where a feature other than a cylindrical shape is located, then the tolerance value represents the distance between two parallel straight lines or planes, or the distance between two uniform boundaries.

❏ Positional tolerances are established with a diameter tolerance zone, as shown in Example 7-2, unless the tolerance zone is between two parallel straight lines or planes, or between two uniform boundaries.

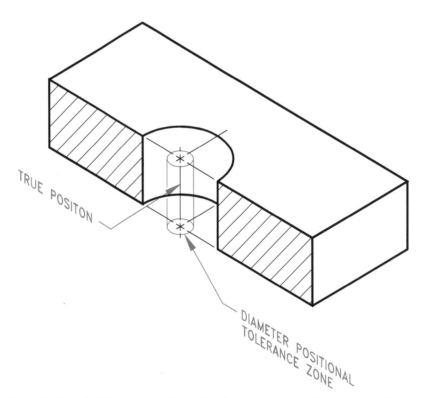

Example 7-2. A diameter positional tolerance zone with true position shown. While this displays the cylindrical tolerance zone, the tolerance zone may also be between two parallel straight lines or planes, or between two uniform boundaries, depending on the application.

A Comparison Between Conventional Tolerancing and Positional Tolerancing

The term *conventional tolerancing* as used in this text refers to the use of conventional coordinate dimensioning practices. A comparison between the use of conventional coordinate location dimensioning and positional dimensioning and tolerancing will help you understand the function of geometric tolerancing for the location of features. The location of a hole using conventional dimensioning and tolerancing methods is shown in Example 7-3.

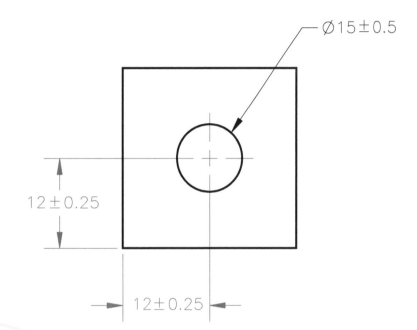

Example 7-3. Location of a hole using conventional coordinate dimensioning system.

The 12±0.25 location dimensions in Example 7-3 establish a total tolerance zone of 0.5. This tolerance zone is square, as you can see in Example 7-4.

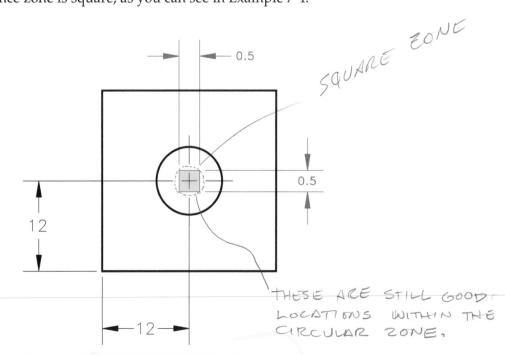

Example 7-4. Conventional location tolerance zone.

The 0.5 square tolerance zone shown in Example 7-5 demonstrates that the actual center of the hole can fall anywhere within the square area and the manufactured part is acceptable.

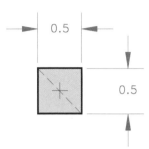

Example 7-5. A close look at the conventional location tolerance zone.

The application of positional tolerancing on the same part allows the acceptable tolerance zone to increase in size. Consider the following points:

❏ The diagonal of the square tolerance zone, represented by the dashed line in Example 7-5, is the greatest distance that allows variation in the location of the center.

❏ The length of this diagonal is equal to a constant of 1.4 times the tolerance of the location dimensions.

❏ Using the tolerance zone from Example 7-5 gives a diagonal length of 1.4 × 0.5 = 0.7. Refer to Example 7-6.

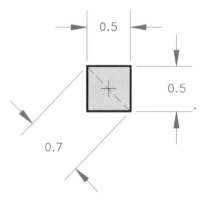

Example 7-6. The length of the diagonal of a square tolerance zone produced by conventional dimensioning.

❏ In a positional tolerance for the location of a hole, this 0.7 diagonal becomes a diameter tolerance zone that is cylindrical in shape through the thickness of the part. This is how a conventional location tolerance may be converted directly to a positional tolerance. It has been proven that the diagonal tolerance zone is acceptable in any direction, thus creating a circular tolerance zone. The result of this action is an increase of 57% in permissible area for the location of the hole. The relationship between the square conventional tolerance and the round positional tolerance zone is shown in Example 7-7. With the use of positional tolerancing, there is an increase of acceptability of mating parts and a possible reduction in manufacturing costs. See Appendix A11 and Appendix A12 for coordinate-to-positional tolerance coversion charts.

57% MORE AREA

See p. 329

Example 7-7. The positional tolerance zone circumscribed about the conventional tolerance zone.

When converting a drawing with conventional location dimensioning to a drawing with positional tolerancing, use the following guide lines:

1. Add datums as appropriate. Notice in Example 7-8 the datums A, B, and C have been placed on the drawing. Perpendicularity of the true position centerline is controlled relative to the primary datum (Datum A in this case). Datum B and Datum C control the location of true position.

2. Change the location dimensions from plus-minus dimensions to basic dimensions, as shown in Example 7-8. This locates the theoretically exact true position of the hole.

3. Add the feature control frame to the diameter dimension, as shown in Example 7-8. Show the positional symbol in the first compartment, followed by the diameter symbol, the calculated positional tolerance, and the MMC symbol unless otherwise specified. MMC is a commonly used material condition symbol with positional tolerancing. However, RFS is assumed unless MMC or LMC is specified. The last three compartments in the feature control frame contain the datum references A, B, and C.

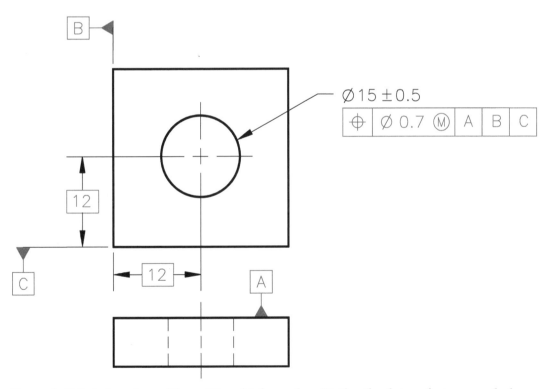

Example 7-8. A drawing with positional tolerancing. Notice the datum feature symbols, basic location dimensions, and feature control frame added.

The previous discussion explained the difference between conventional tolerancing and positional tolerancing, and how to make a direct conversion of a drawing with conventional dimensioning to positional tolerancing. In many situations, the engineer gives the desired positional tolerances. Only direct application to the drawing is required and no conversion is necessary.

When locating holes using positional tolerancing, the location dimensions must be basic. This may be accomplished by applying the basic dimension symbol to each of the basic dimensions, or specifying on the drawing, or in a reference document, the general note "UNTOLERANCED DIMENSIONS LOCATING TRUE POSITION ARE BASIC."

Positional tolerances are often applied at MMC. However, either MMC or LMC must be indicated in the feature control frame to the right of the positional tolerance as applicable. Otherwise, RFS is assumed. Example 7-9 shows the cylindrical tolerance zone that is established by the positional tolerance specified in Example 7-8. The true position centerline is perpendicular to the primary datum. The centerline of the hole can be anywhere within the diameter and length of the specified cylindrical positional tolerance zone at the maximum material condition size of the hole.

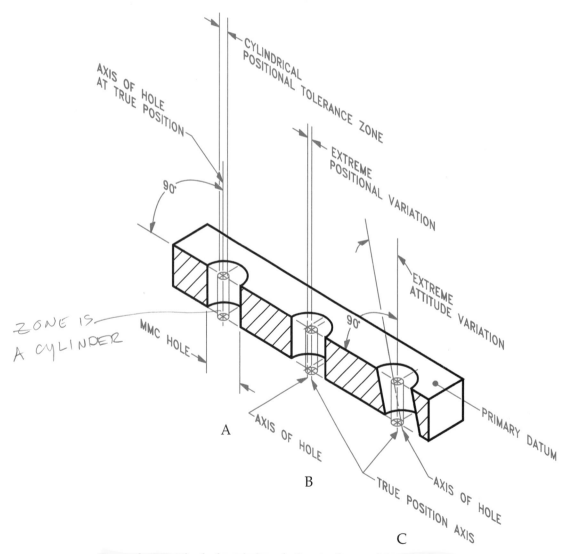

**Example 7-9. The hole axis in relation to the positional tolerance zone.
A—Shows the axis of the hole at true position. B—Shows the axis of the hole
at extreme side of the positional tolerance zone. This is called extreme
positional variation. C—Shows the axis of the hole at an extreme angle inside
the positional tolerance zone. This is called extreme attitude variation.**

As you look at Example 7-9 notice that the hole in part "A" shows the axis of the hole at true position. The example in part "B" shows the axis of the hole at the extreme side of the positional tolerance zone. This is referred to as *extreme positional variation.* The example in part "C" shows the axis of the hole at an extreme angle inside the positional tolerance zone. This is referred to as *extreme attitude variation.*

Positional Tolerance at MMC

The maximum material condition (MMC) of a feature means that the actual size contains the maximum amount of material permitted by the size dimension tolerance for that feature. A hole or other internal feature is at MMC when the actual size is at the lower limit. A shaft or other external feature is at MMC when the actual size is at the upper limit.

A positional tolerance at MMC means that the specified positional tolerance applies when the feature is manufactured at MMC. The axis of a hole must fall within a cylindrical tolerance zone with an axis located at true position. The diameter of this cylindrical tolerance zone is equal to the specified positional tolerance when the hole is manufactured at MMC. The positional tolerance is then allowed to increase equal to the amount of change or departure from MMC. The maximum amount of positional tolerance is when the feature is produced at LMC, as shown in the analysis provided in Example 7-10. When MMC is applied to a positional tolerance, the following formulas may be used to calculate the positional tolerance at any produced size:

Internal Feature (as in Example 7-10):
Actual Size – MMC + Specified Positional Tolerance = Applied Positional Tolerance

External Feature:
MMC – Actual Size + Specified Positional Tolerance = Applied Positional Tolerance

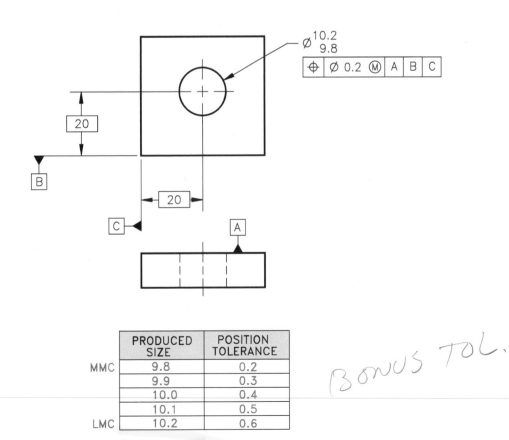

	PRODUCED SIZE	POSITION TOLERANCE
MMC	9.8	0.2
	9.9	0.3
	10.0	0.4
	10.1	0.5
LMC	10.2	0.6

Example 7-10. Positional tolerance applied in regard to MMC.

Positional Tolerance Based on the Surface of a Hole

Positional tolerance applied at MMC may also be explained in regard to the surface of the hole rather than the hole axis. In this explanation, all elements of the hole surface must be inside a theoretical boundary located at true position and the hole must be produced within the specified size limits, as shown in Example 7-11.

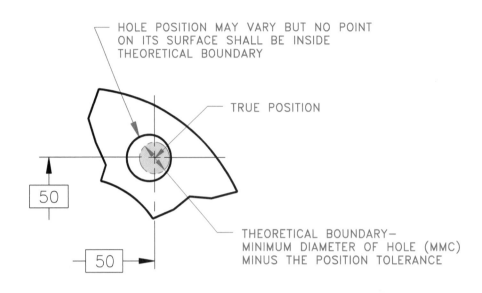

HOLE POSITION MAY VARY BUT NO POINT ON ITS SURFACE SHALL BE INSIDE THEORETICAL BOUNDARY

TRUE POSITION

THEORETICAL BOUNDARY– MINIMUM DIAMETER OF HOLE (MMC) MINUS THE POSITION TOLERANCE

MMC HOLE – POSITIONAL TOLERANCE = THEORETICAL BOUNDARY

Example 7-11. The boundary for the surface of a hole at MMC.

Zero Positional Tolerancing at MMC

Zero geometric tolerancing was introduced in Chapter 6. This concept may also be applied to positional tolerances. You have already seen that the application of positional tolerance at MMC allows the positional tolerance zone to exceed the amount specified when the feature is produced at any actual size other than MMC. Zero positional tolerance may be specified when it is important to be certain that the tolerance is totally dependent on the actual size of the feature. When this is done, the positional tolerance is zero when the feature is produced at MMC and must be located at true position. When the actual size of the feature departs from MMC, then the positional tolerance is allowed to increase equal to the amount of departure. The total allowable variation in positional tolerance is at LMC, unless a maximum tolerance is specified. Other than specifying zero positional tolerance at MMC in the feature control frame, this is the same application explained in the previous discussion "Positional Tolerance at MMC." When zero positional tolerance at MMC is specified, the engineer normally applies the MMC of the hole at the absolute minimum required for insertion of a fastener when located at true position. Refer to Example 7-12.

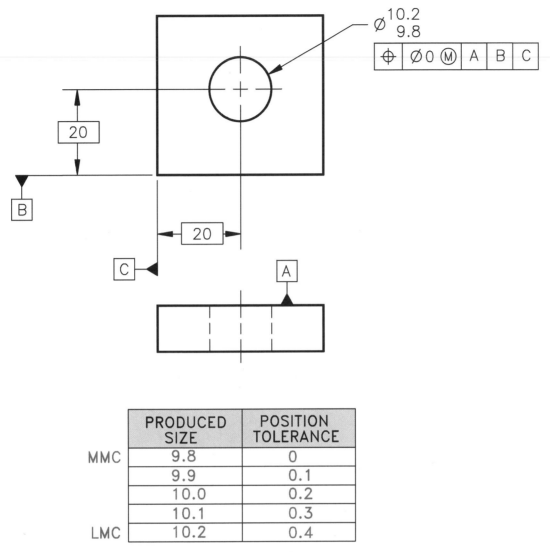

	PRODUCED SIZE	POSITION TOLERANCE
MMC	9.8	0
	9.9	0.1
	10.0	0.2
	10.1	0.3
LMC	10.2	0.4

Example 7-12. Zero positional tolerance at MMC.

Positional Tolerance at RFS

Regardless of feature size (RFS) is assumed when no material condition symbol is placed after the positional tolerance in the feature control frame. RFS may be applied to the positional tolerance when it is desirable to maintain the given positional tolerance at any produced size. The use of RFS may also be applied to the datum reference that must be maintained regardless of the actual feature sizes. The application of RFS requires closer controls of the features involved because the size of the positional tolerance zone may not get larger as when MMC is used. Remember, RFS is assumed for the geometric tolerance and datum reference unless otherwise specified. Refer to Example 7-13.

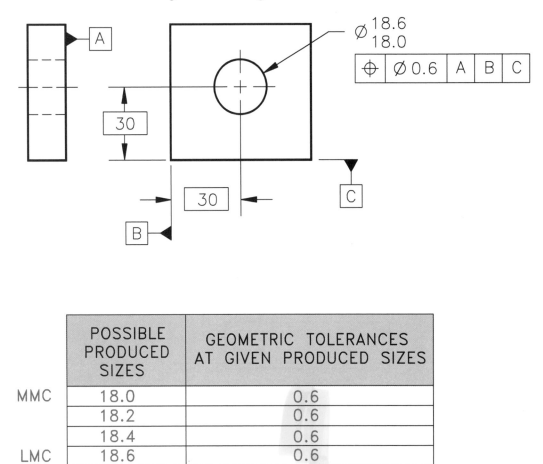

POSSIBLE PRODUCED SIZES	GEOMETRIC TOLERANCES AT GIVEN PRODUCED SIZES
18.0	0.6
18.2	0.6
18.4	0.6
18.6	0.6

MMC — 18.0
LMC — 18.6

Example 7-13. Positional tolerance at RFS.

Positional Tolerance at LMC

Positional tolerance at least material condition (LMC) is used to control the relationship of the feature surface and the true position of the largest hole size. The function of LMC is sometimes used to control minimum edge distance. When the LMC material condition symbol is used in the feature control frame, the given positional tolerance is held at the LMC produced size. Then as the produced size departs from LMC toward MMC, the positional tolerance increases equal to the amount of change from LMC. The maximum amount of positional tolerance is applied at the MMC produced size, as shown in Example 7-14. When using the LMC control, perfect form is required at the LMC produced size. LMC specifications are limited to positional tolerances where the use of MMC does not give the desired control and RFS is too restrictive.

When LMC is applied to a positional tolerance the following formulas may be used to calculate the positional tolerance at any produced size:

Internal Feature:
LMC – Actual Size + Specified Positional Tolerance = Applied Positional Tolerance

External Feature:
Actual Size – LMC + Specified Positional Tolerance = Applied Positional Tolerance

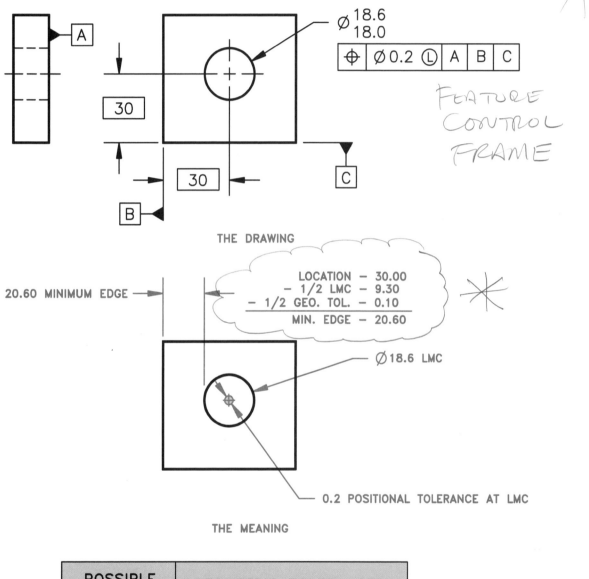

Example 7-14. Positional tolerance at LMC.

	POSSIBLE PRODUCED SIZES	GEOMETRIC TOLERANCES AT GIVEN PRODUCED SIZES
MMC	18.0	0.8
	18.2	0.6
	18.4	0.4
LMC	18.6	0.2

LOCATING MULTIPLE FEATURES

Multiple features of an object can be dimensioned using positional tolerancing. When this is done, the location of the features must be dimensioned from datums and between features using datum or chain line dimensioning related to rectangular or polar coordinates.

Rectangular coordinate dimensioning is where linear dimensions are used to locate features from planes, centerlines, or center planes. Refer to Example 7-15.

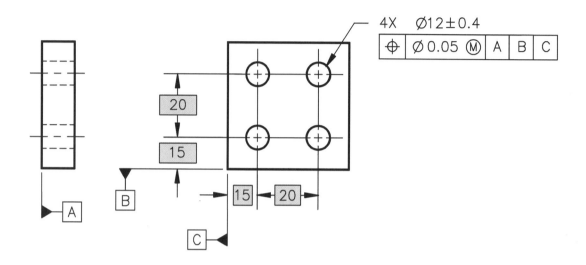

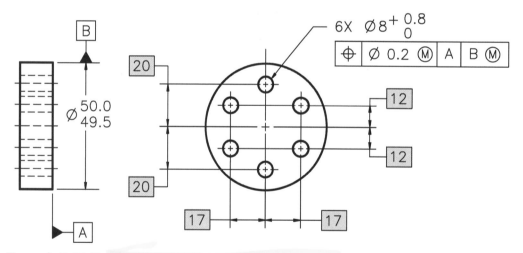

Example 7-15. Rectangular coordinate dimensioning.

Polar coordinate dimensioning is where angular dimensions are combined with other dimensions to locate features from planes, centerlines, or center planes. Refer to Example 7-16.

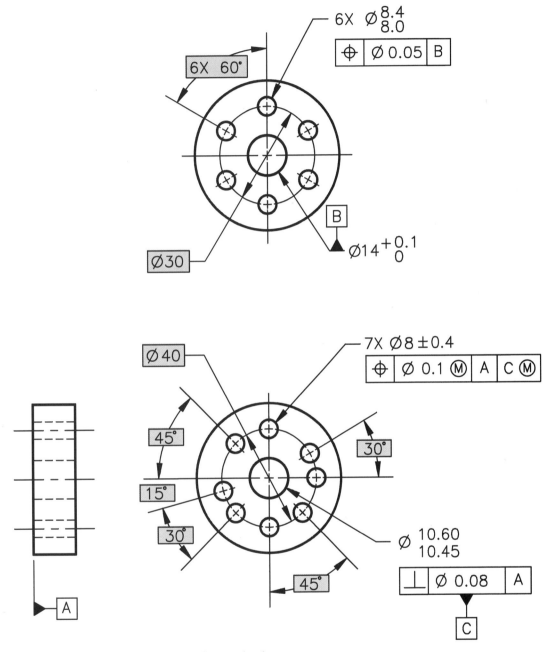

Example 7-16. Polar coordinate dimensioning.

When multiple features are located, the following guidelines apply.

❑   The pattern of features is located collectively in relationship to datum features that are not subject to size changes. The actual centers of all features in the pattern must lie on, or be within, the specified positional tolerance zone when measured from the given datums.

❑   Multiple patterns of features are considered a single composite pattern if the related feature control frames have the same datums, in the same order of precedence, with the same material condition symbols.

Locating a Single Composite Pattern

A group of features is referred to as a *single composite pattern* when they are located relative to common datum features not subject to size tolerance, or to common datum features of size specified on an RFS basis. All of the location dimensions are basic from a common datum reference frame as previously discussed. All of the holes can be checked together, as shown in Example 7-17.

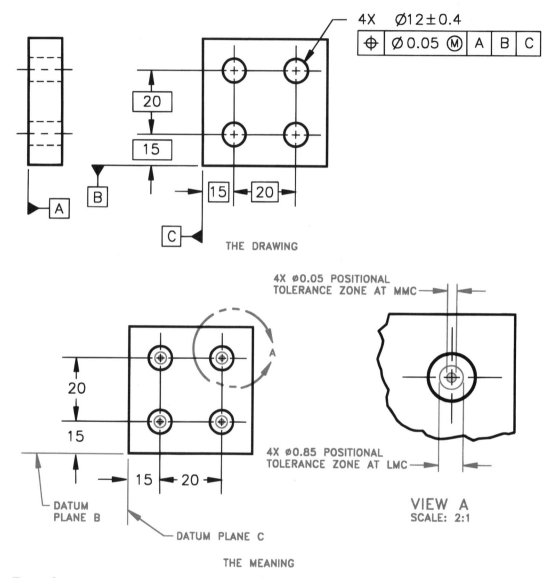

Example 7-17. Locating a single composite pattern.

Locating Features in a Pattern with Separate Requirements

A pattern of features is located as a single composite pattern, as previously discussed, when there is no note given below the feature control frame that specifies otherwise. The note "SEP REQT" (separate requirement) is placed below each feature control frame when it is desired to have some features in the pattern specified with geometric tolerance that are separate or different from the pattern. This allows features within the total pattern to be treated as separate patterns and to have their own datum reference frame. This may be done when a pair of features in a pattern are different in size or have different location requirements than the other features, as shown in Example 7-18.

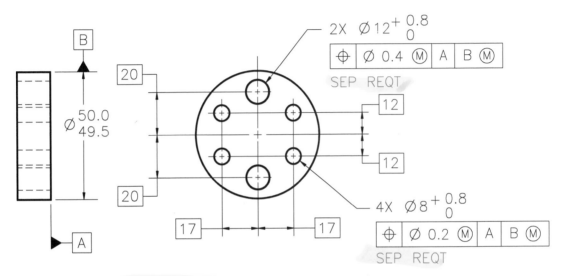

Example 7-18. Locating features in a pattern with separate requirements.

Composite Positional Tolerance

Composite positional tolerancing is used when it is desirable to permit the location of a pattern of features to vary within a larger tolerance than the positional tolerance specified for each feature. For this application, the feature control frame is doubled in height and divided into two parts. Only one positional geometric characteristic symbol is used and it is placed in one double height feature control frame compartment. The upper part of the feature control frame is the *pattern-locating control* and specifies the larger positional tolerance for the pattern of features as a group. The lower entry is called the *feature-relating control* and specifies the smaller positional tolerance for the individual features within the pattern. The pattern-locating control is located first to related datums with basic dimensions. The feature-relating control is free to shift and slant within the boundaries established by the pattern-locating control. Only the primary datum is given to control orientation (perpendicularity) in the feature-relating control. The tolerance zone of an individual feature may extend partly beyond the group zone, but the feature axis must fall within the confines of both zones, as shown in Example 7-19.

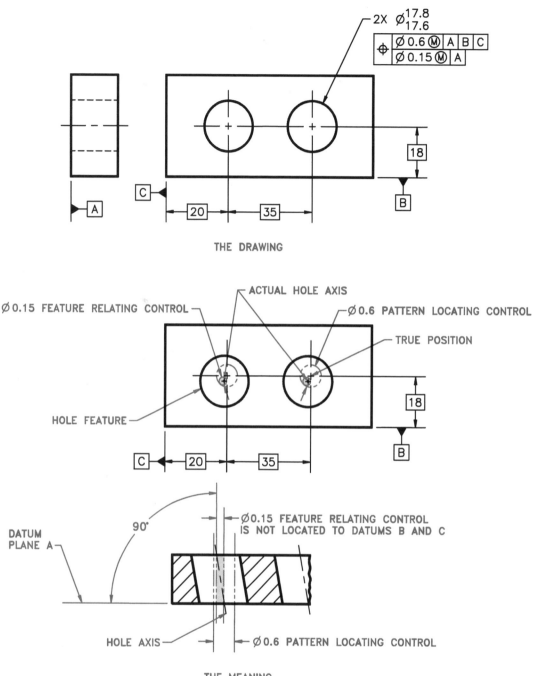

Example 7-19. Composite positional tolerancing.

The composite positional tolerance may also be applied with two datum references placed in the lower half of the feature control frame. When this is done, the feature-relating control must remain as a group that is perpendicular to the primary datum and parallel to the secondary datum. Notice in Example 7-20 how the pattern-locating zones and the feature-relating zones are parallel. This feature-relating orientation to the secondary datum is not required when only the primary datum reference is placed in the lower half of the feature control frame, as in Example 7-19.

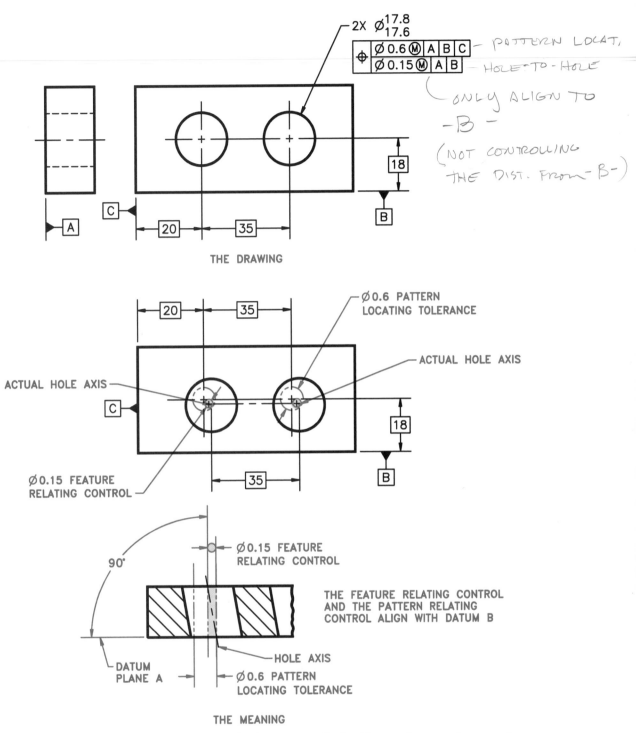

THE DRAWING

THE MEANING

Example 7-20. The composite positional tolerance applied with two datum references in the lower half of the feature control frame. This makes the pattern-locating zone and the feature-relating zone parallel.

Two single-segment feature control frames

The composite positional tolerance is specified by a feature control frame doubled in height with one positional symbol shown in the first compartment. Also provided is a single datum reference given for orientation, or a double datum reference given for orientation and alignment with respect to the feature-relating control. The *two single-segment feature control frame* is similar, except there are two positional symbols, each displayed in a separate compartment, and a two datum reference in the lower half of the feature control frame, as shown in Example 7-21. The top feature control frame is the pattern-locating control and works as previously discussed. The lower feature control frame is the feature-relating control where two datums control the orientation and the alignment with the pattern-locating control. This type of positional tolerance provides a tighter relationship of the holes within the pattern than the composite positional tolerance, because both the pattern-locating zones and the feature-relating zones must remain the same distance from the secondary datum. Refer to Example 7-21.

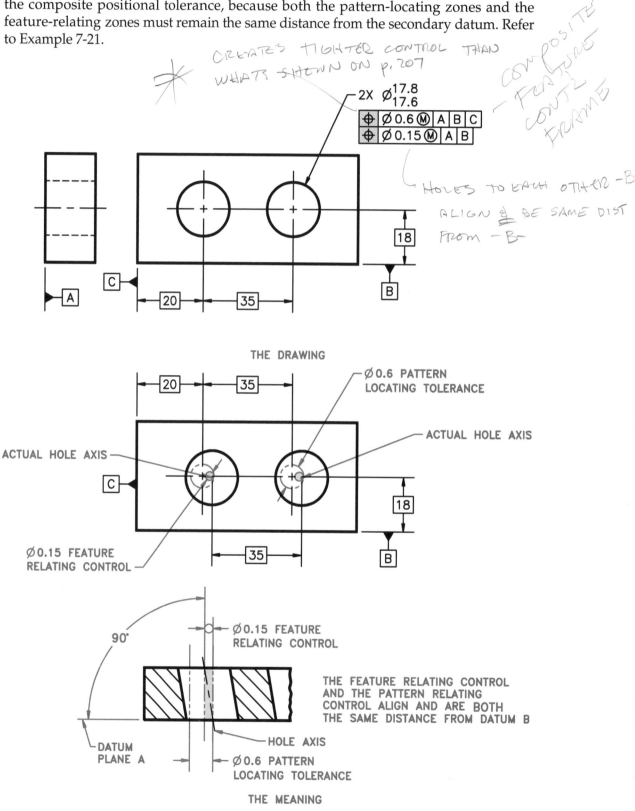

Example 7-21. The two single-segment positional tolerance application.

Composite positional tolerance applied to circular patterns

Composite positional tolerancing may be applied to circular patterns of features. In this application, the pattern-locating tolerance zones are located using a basic diameter and basic angle between features and oriented to the specified datum reference frame. The feature-relating tolerance zones are held perpendicular to the primary datum, controlled as a group by the basic dimensions, and are partially or totally within the boundaries of the pattern-locating tolerance zones. The actual feature axes must fall within the bounds of both tolerance zones. Refer to Example 7-22.

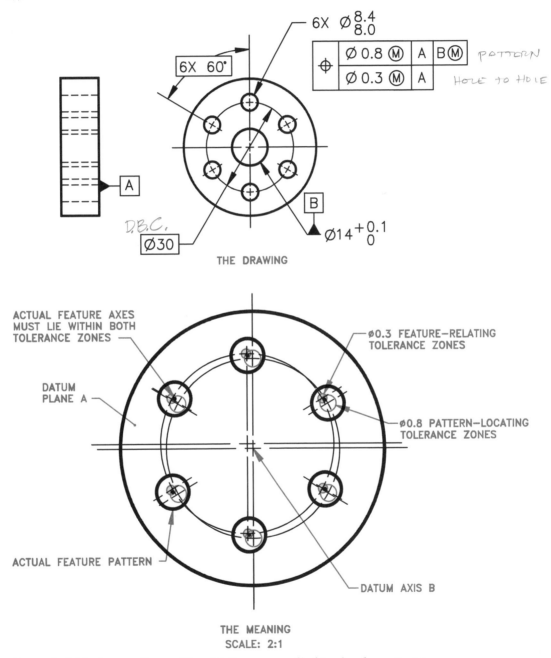

Example 7-22. Composite positional tolerance applied to circular patterns.

Two single-segment feature controls may also be applied to circular patterns. The positional geometric characteristic symbol is displayed twice as previously discussed. This is used when it is necessary for the pattern-locating zones and the feature-relating zones to be located from a common datum axis. Notice in Example 7-23 that Datum B (the datum axis) is listed in both halves of the feature control frame.

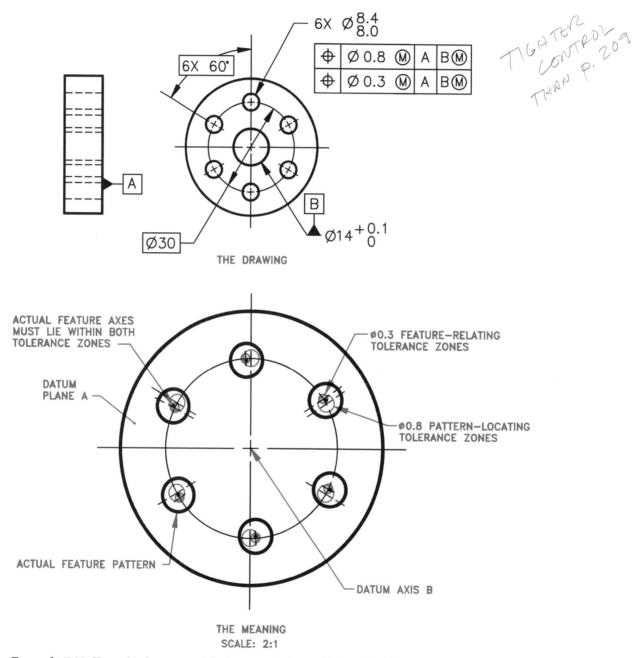

Example 7-23. Two single-segment feature controls applied to circular patterns.

POSITIONAL TOLERANCE OF COAXIAL FEATURES

Coaxial features are those features having a common axis such as counterbores, countersinks, and counterdrills. When the positional tolerance of the coaxial features is to be the same (for example, the same for the hole and the associated counterbore) then the feature control frame is placed below the note specifying the hole and counterbore, as shown in Example 7-24. When this is done, the positional tolerance zone diameter is identical for the hole and counterbore relative to the specified datums.

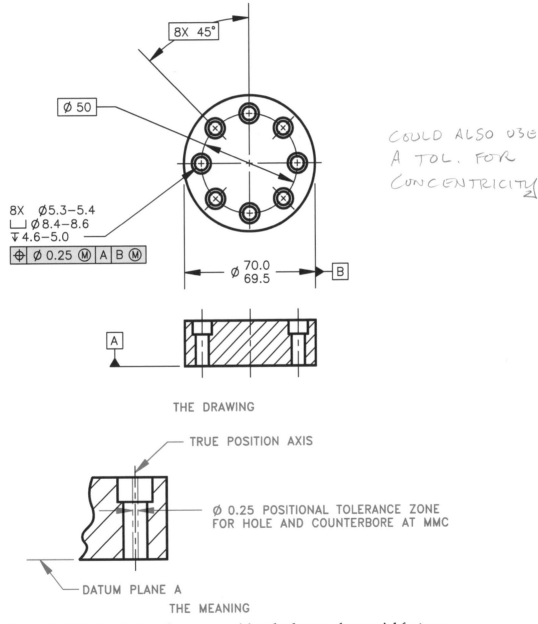

Example 7-24. Displaying the same positional tolerance for coaxial features.

Different positional tolerances may be applied to coaxial features related to the same datum features. For example, when the positional tolerance is different for the counterbore and the related hole, then one feature control frame is placed under the note specifying the hole size and another feature control frame placed under the note specifying the counterbore. Refer to Example 7-25. This may be possible when the counterbore is a different tolerance than the hole.

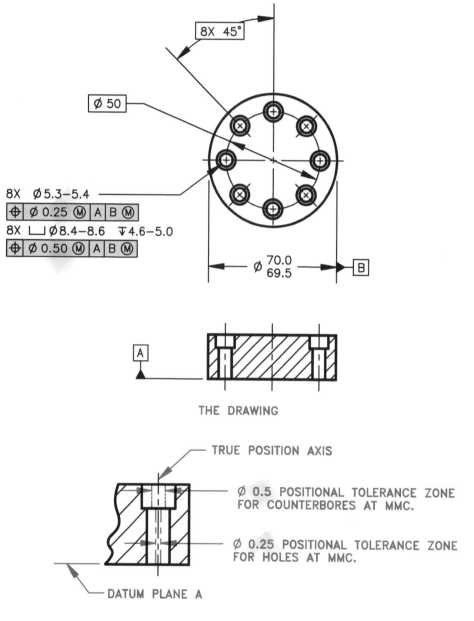

THE DRAWING

THE MEANING

Example 7-25. Displaying different positional tolerances for coaxial features.

Positional tolerances may also be applied to coaxial features such as counterbore holes by controlling individual counterbore-to-hole relationships relative to different datum features. The application is similar to the one shown and detailed in Example 7-25. However, an additional note is placed under the datum feature symbol for the hole and under the feature control frame for the counterbore indicating the number of places each applies on an individual basis, as shown in Example 7-26. With this method, the holes are located as a single composite pattern and then the counterbores are located individually to each related hole, with the axis of the hole as the aligning datum.

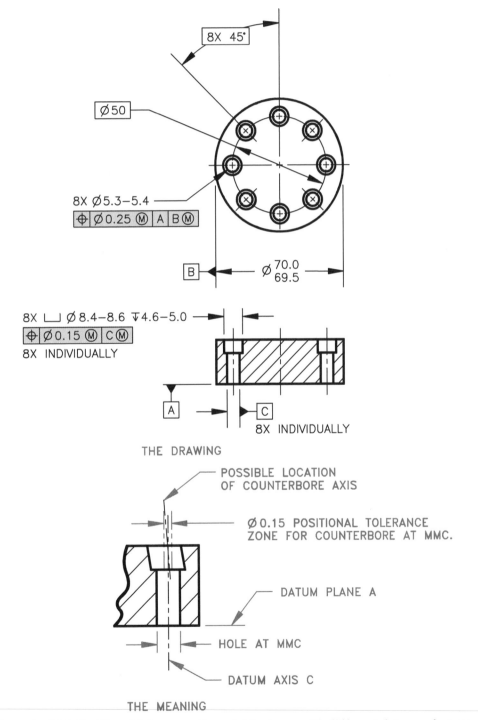

Example 7-26. Positional tolerance of coaxial features with different datum reference.

Coaxial Positional Tolerance

Coaxial features are two or more features that lie on a common axis. In the previous discussion, counterbore features were described as being coaxial. This discussion is in regard to coaxial features that are established as holes that are apart, but in alignment, as shown in Example 7-27. A *coaxial positional tolerance* may be used to control the alignment of two or more holes that share a common axis. This is used when a tolerance of location alone does not provide the necessary control of alignment of the holes and a separate requirement must be specified. When this is done, the positional tolerance feature control frame is doubled in height. The top half of the frame is used to specify the coaxial diameter tolerance zones at MMC located at true position relative to the specified datums where the axes of the holes, as a group, must lie. The lower half of the feature control frame is used to designate the coaxial diameter tolerance zones at MMC where the axes of the holes must lie relative to each other. Refer to Example 7-27.

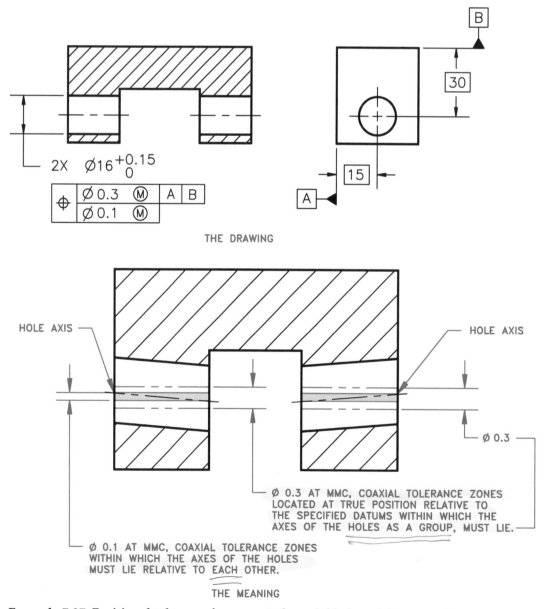

Example 7-27. Positional tolerance for separated coaxial holes of the same size.

When the positional tolerance of coaxial holes of different sizes are to be manufactured then the drawing looks similar to Example 7-23, except the different sized holes are dimensioned independently and the note "TWO COAXIAL HOLES" (two is the number of holes) is placed below the feature control frame. This acknowledges that the same positional tolerance zone requirements apply to all holes.

POSITIONAL TOLERANCE OF NONPARALLEL HOLES

Positional tolerances may be used in situations where the axes of the holes are not parallel to each other and where they may also be at an angle to the surface, as shown in Example 7-28.

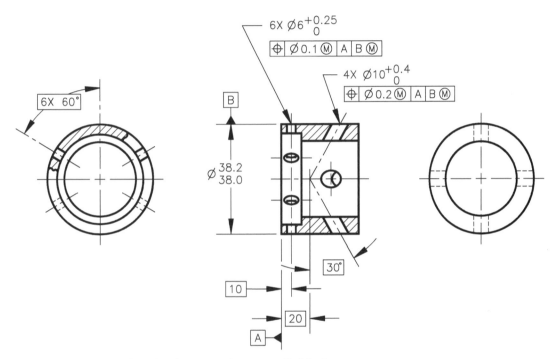

Example 7-28. Positional tolerance of nonparallel holes.

LOCATING SLOTTED HOLES

Slotted holes may be located to their centers with basic dimensions from established datums. When a greater positional tolerance is placed on the length than the width, the feature control frame is added to both the length and width dimensions of the slotted features. The positional tolerance may also be the same for the length and the width. In this case, the feature control frame is separated from the size dimension and is connected to the feature with a leader. In either case, the word "BOUNDARY" is placed below the feature control frame, as shown in Example 7-29. This means that each slotted feature is controlled by a theoretical boundary of identical shape that is located at true position. The size of each slot must remain within the size limits and no portion of the slot surface may enter the theoretical boundary. The length and width of the boundary is calculated with the appropriate formula from the following:

MMC length – positional tolerance = boundary length

MMC width – positional tolerance = boundary width

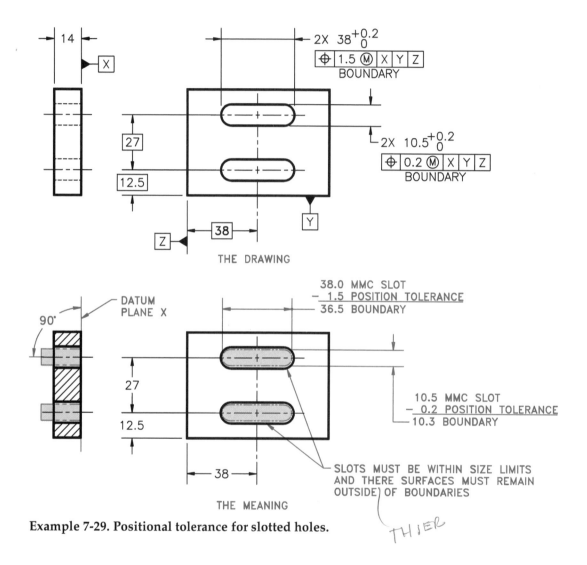

Example 7-29. Positional tolerance for slotted holes.

POSITIONAL TOLERANCE OF SPHERICAL FEATURES

A positional tolerance may be used to control the location of a spherical feature relative to other features of a part. When dimensioning spherical features, the spherical diameter symbol precedes the feature size dimension. The feature control frame is placed below the size dimension and the positional tolerance zone is spherical in shape, as shown in Example 7-30.

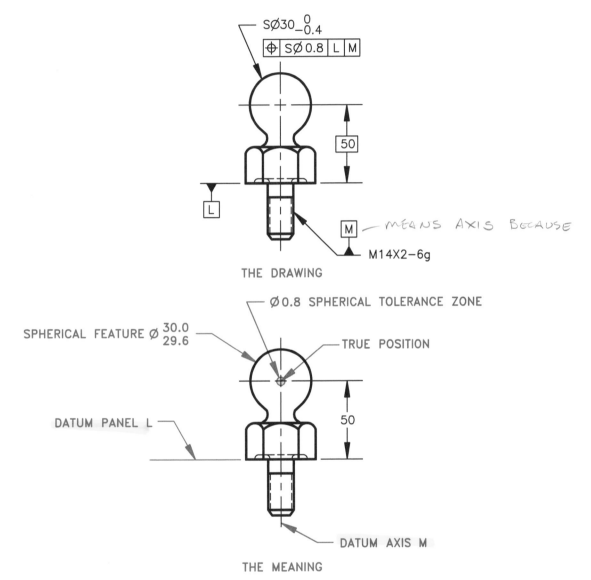

Example 7-30. Positional tolerance of a spherical feature.

Name: _____

1. Describe the purpose of location tolerances. _____

    _____

    _____

2. _____ is used to define a zone where the center, axis, or center plane of a feature of size is permitted to vary from true position.

3. Define true position. _____

    _____

4. _____ dimensions are used to establish true position from specified datum features and between interrelated features.

5. Location tolerancing is specified by a(n) _____ or a(n)

    _____ symbol, a(n) _____ and appropriate

    _____ references placed in a feature control frame.

6. When positional tolerancing is used, the MMC or LMC material condition symbols must be specified after the tolerance and applicable datum reference in the feature control frame. Otherwise, RFS is assumed. True or False.

7. Redraw the object below making a complete conversion from the conventional coordinate dimensioned drawing shown to a positional toleranced drawing as follows:

    a) Calculate the positional tolerance required to convert this drawing to a positional tolerance drawing. (Show your calculations.)

    b) Use full scale.

    c) Use proper size symbols.

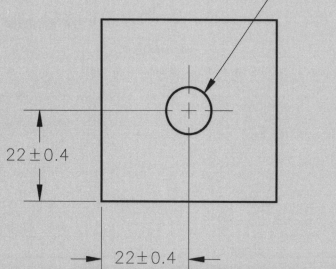

Use the space below for the drawing in question 7.

Show your calculations for question 7 here:

_____

_____

_____

_____

_____

8. When locating features using positional tolerancing, the basic dimensions may be drawn by placing the basic dimension symbol around each basic dimension, or specifying on the drawing (or in a reference document) the following general note:

_____

_____

9. Given the following drawing, a reference chart showing a range of possible produced sizes, and four optional feature control frames that may be applied to the diameter dimension, provide the positional tolerance at each possible produced size for each feature control frame application.

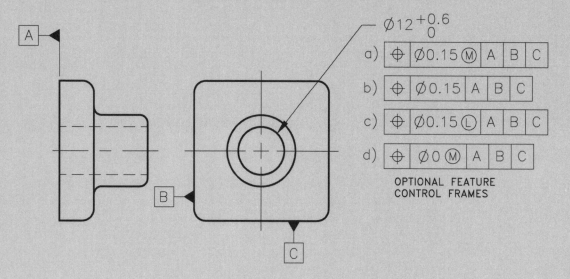

DIAMETER TOLERANCE ZONES ALLOWED

Possible Produced Sizes	a) MMC	b) RFS	c) LMC	d) 0 at MMC
12.0	_____	_____	_____	_____
12.2	_____	_____	_____	_____
12.4	_____	_____	_____	_____
12.6	_____	_____	_____	_____

10. Name the two types of dimensioning systems that are normally used when locating multiple features. _____

_____

_____

_____

11. The positional tolerancing of slotted holes is accomplished by locating basic dimensions to the centers of the slots from datums, and adding the feature control frame to the length and width dimension. True or False.

12. Describe when composite positional tolerancing is used. _____

_____

_____

_____

13. When using composite positional tolerancing, the upper part of the feature control frame is referred to as the _____ and specifies the larger tolerance for the pattern of features as a group, while the lower half of the frame is called the _____ and specifies a smaller positional tolerance for individual features within the pattern.

14. Positional tolerances may be used in situations where the axes of the holes are not parallel to each other and where they may be at an angle to the surface. True or False.

15. What is the shape of the positional tolerance zone when using positional tolerancing to control the location of a spherical feature? _____

    _____

16. When the axis of a hole is at the extreme side of a positional tolerance zone, it is referred to as _____

    _____.

17. When the axis of a hole is at an extreme angle inside the positional tolerance zone, it is referred to as _____

    _____.

18. Give the formulas, for internal and external features, that can be used when calculating the positional tolerance at any produced size when MMC is applied to the positional tolerance:

    Internal feature:

    _____

    _____

    External feature:

    _____

    _____

19. Give the formulas, for internal and external features, that can be used when calculating the positional tolerance at any produced size when LMC is applied to the positional tolerance:

    Internal feature:

    _____

    _____

    External feature:

    _____

    _____

20. Calculate the minimum edge distance between the edge of a hole and the out-
    side surface of the part for an LMC positional tolerance application given the
    following information: (dimensions are in millimeters)

    Location dimension: 40.5

    Positional tolerance: 0.4

    MMC of hole: 12.5

    LMC of hole: 13.5

    Calculations:

    _____

    _____

    _____

    _____

21. What is the purpose of the note "SEP REQT" placed below the feature control
    frame for locating features in a pattern? _____

    _____

    _____

    _____

    _____

    _____

22. Why is the word "BOUNDARY" placed below the feature control frame when
    locating slotted features? _____

    _____

    _____

    _____

    _____

    _____

23. Give the formula for calculating the slot boundary discussed in question 22: ___

    _____

    _____

24. How is the feature control frame placed in relation to a hole and counterbore
    when the positional tolerance is the same for the hole and counterbore? _____

    _____

    _____

25. How is the feature control frame placed in relation to a hole and counterbore
    when the positional tolerance is different for the hole and counterbore? _____

    _____

    _____

26. What is the difference between the appearance of the feature control frame used for a composite positional tolerance and the one used for a two single-segment positional tolerance?

_____

_____

_____

_____

_____

_____

27. Explain the primary difference between the composite positional tolerance and the two single-segment positional tolerance. _____

_____

_____

_____

_____

_____

_____

_____

_____

_____

28. Explain the primary difference between the composite positional tolerance and the two single-segment positional tolerance as applied to circular patterns.

_____

_____

_____

_____

_____

_____

_____

_____

Print Reading Exercise

Name: _____

7

The following print reading exercise uses actual industry prints with related questions that require you to read specific dimensioning and geometric tolerancing representations. The answers should be based on previously learned content of this book. The prints used are based on ASME standards, however company standards may differ slightly. When reading these prints, or any other industry prints, a degree of flexibility may be required to determine how individual applications correlate with the ASME standards.

Print Reading Exercise

Refer to the print of the BRACKET found on page 304.

1. Refer to the Ø.875±.005 dimension:

    a) In regard to the positional tolerance, what is the location called that is 1.500 basic from Datum B and 1.950 basic from Datum C? _____

    b) What is the positional tolerance? _____

    c) Describe Datum D. _____

    _____

    _____

    _____

    d) Name the location dimensions with boxes around and describe their purpose.

    _____

    _____

    _____

    e) Give the positional tolerance at the following actual produced sizes:

Produced Sizes	Positional Tolerance
.880	_____
.878	_____
.876	_____
.874	_____
.872	_____
.870	_____

2. Refer to the Ø.187±.003 dimension:

    a) How many of these features are there?_____

    b) Are the location dimensions placed using rectangular or polar coordinate dimensioning? _____

    c) Which of the following most closely describes the positional tolerance applied to these features:

        1) Single composite pattern.

        2) Pattern with separate requirements.

        3) Composite positional tolerance.

        4) Two single-segment feature control frame.

3. Refer to the Ø.156±.005 dimension:

    a) How many of these features are there?_____

    b) Are the location dimensions placed using rectangular or polar coordinate dimensioning? _____

    c) Which of the following most closely describes the positional tolerance applied to these features:

        1) Single composite pattern.

        2) Pattern with separate requirements.

        3) Composite positional tolerance.

        4) Two single-segment feature control frame.

    d) Completely describe the meaning of the information provided in the feature control frame. _____

        _____

        _____

        _____

        _____

        _____

        _____

        _____

        _____

        _____

        _____

        _____

        _____

        _____

        _____

        _____

4. Refer to the ∅.437±.005 dimension:

   a) How many of these features are there? _____

   b) Which of the following most closely describes the positional tolerance applied to these features:

      1) Single composite pattern.

      2) Pattern with separate requirements.

      3) Composite positional tolerance.

      4) Two single-segment feature control frame.

   d) Completely describe the meaning of the information provided in the feature control frame. _____

   _____

   _____

   _____

   _____

   _____

   _____

   _____

   _____

   _____

   _____

   _____

   _____

   _____

   _____

   _____

Refer to the print of the MOUNTING PLATE (UPPER)-FRAME ASSY 3 AXIS HP found on page 307.

5. Refer to the ∅.188 THRU COUNTERBORE ∅.313 DEPTH 1.070 specification:

   a) Why is there a separate feature control frame for the hole and counterbore?

   _____

   _____

   _____

   _____

b)   Provide the positional tolerance at each of the following possible produced sizes for the hole and counterbore:

Hole		Counterbore	
Produced Sizes	Positional Tolerance	Produced Sizes	Positional Tolerance
.191	_____	.316	_____
.190	_____	.314	_____
.188	_____	.313	_____
.187	_____	.312	_____

Refer to the print of the DOUBLE V-BLOCK found on page 310.

6. Refer to the Ø12.70/12.65 dimension.

a)   How is it possible to have a zero positional tolerance? _____

_____

_____

_____

b)   Provide the positional tolerance at the following possible produced sizes:

Produced Sizes	Positional Tolerance
12.70	_____
12.69	_____
12.68	_____
12.67	_____
12.66	_____
12.65	_____

7. Describe Datum D. _____

_____

_____

8. Refer to the 44.45 dimension.

a)   What is the positional tolerance? _____

b)   What is the datum reference? _____

c)   Explain the meaning of the information in the feature control frame.

_____

_____

_____

_____

_____

_____

_____

Refer to the print of the PLATE-TOP MOUNTING found on page 311.

9. Refer to the ⌀.290 THRU COUNTERBORE ⌀.625 DEPTH SHOWN
    specification:

    a) Are these features located using rectangular or polar coordinate dimensioning?

    _____

    b) How many of these features are there? _____

    c) Why is there the same feature control frame for the hole and counterbore?

    _____

    _____

    _____

Chapter 8

Location Tolerances (Part II) and Virtual Condition

This chapter continues the discussion on location tolerances with emphasis on the following applications:

❏ Positional tolerances to mating parts.
❏ The use of projected tolerance zones.
❏ Virtual condition.
❏ Positional tolerancing for coaxiality.
❏ Concentricity.
❏ Symmetry.
❏ Positional tolerancing for symmetrical features.

FASTENERS

The application of geometric tolerancing, such as orientation or location tolerances, to threaded fasteners is applied to the axis of a cylinder established by the *pitch diameter* of the thread. Example 8-1 shows the elements of an external screw thread. Example 8-2 shows the elements of an internal screw thread.

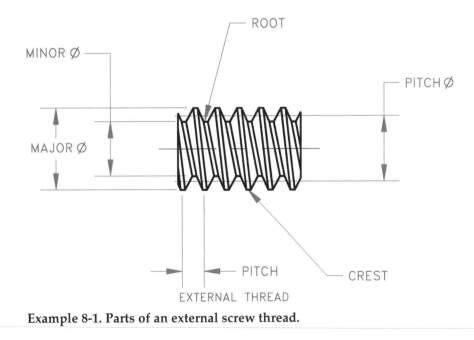

Example 8-1. Parts of an external screw thread.

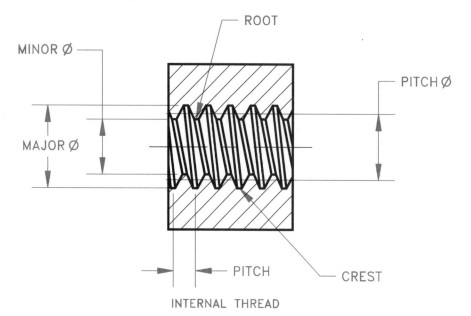

Example 8-2. Parts of an internal screw thread.

If you want the geometric tolerance of the screw thread to be applied to the major diameter or the minor diameter, rather than the pitch diameter, then you need to place the note "MAJOR DIA" or "MINOR DIA" (as appropriate) below the related feature control frame or datum feature symbol.

When the geometric tolerance and datum reference or datum feature is established for gears and splines the specific feature of the gear or spline must be noted below the feature control frame or datum feature symbol. The options include "MAJOR DIA," "PITCH DIA," or "MINOR DIA."

Floating Fasteners

The term *floating fastener* applies to a situation where two or more parts are assembled with fasteners such as bolts and nuts, and all parts have clearance holes for the bolts. A floating fastener situation is shown in Example 8-3. Notice Parts "A" and "B" are fastened together by a bolt, and a nut is required to hold the parts secure. When the holes in a pattern are the same diameters, the bolts used are the same diameters, and the same positional tolerance for all holes is to be the same, then the positional tolerance may be calculated using the formula:

$$\text{MMC HOLE} - \text{MMC FASTENER (BOLT)} = \frac{\text{POSITIONAL TOLERANCE}}{\text{FOR EACH PART}}$$

Note that each part is calculated separately.

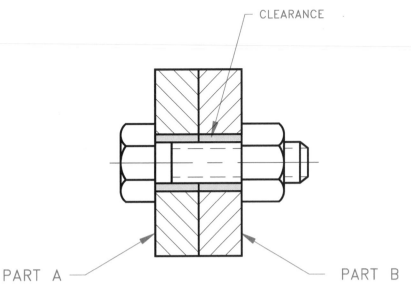

Example 8-3. The floating fastener.

Given a situation where an M12×1.5 bolt is used to fasten together two identical parts with a hole diameter of 13.0/12.5, the positional tolerance required can be calculated as follows:

MMC HOLE (12.5) – MMC BOLT (12) = POSITIONAL TOLERANCE (0.5)

The MMC of a bolt is considered to be the nominal size, which is the same as the major diameter. The major diameter of the M12×1.5 thread is 12 millimeters. The resulting drawing is shown in Example 8-4.

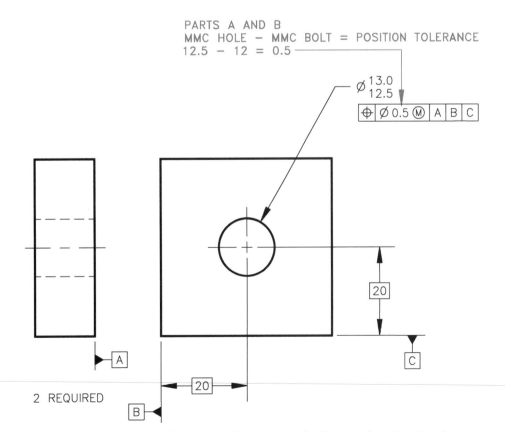

Example 8-4. Calculating and showing the positional tolerance for a floating fastener.

Fixed Fasteners

The term *fixed fastener* applies to a situation where one of the parts to be assembled has a held-in-place fastener such as a threaded hole for a bolt, screw, or stud. This applies to all holes of the same size in a pattern where the same positional tolerance is specified. An example of a fixed fastener situation is shown in Example 8-5.

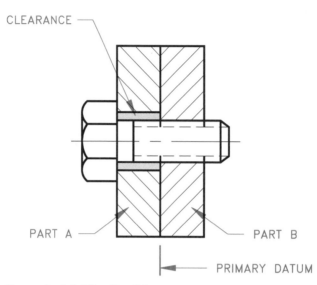

Example 8-5. The fixed fastener.

Notice in Example 8-5 that Part "A" has a clearance hole and Part "B" is threaded. Part "B" acts as a fastener, much like a nut. Therefore, a nut is not required as in the floating fastener situation. Notice that only Part "A" has clearance around the fastener. This means that half as much positional tolerance is applied as compared to a floating fastener. The fixed fastener positional tolerance may be calculated using the formula:

$$\frac{\text{MMC HOLE} - \text{MMC FASTENER (BOLT)}}{2} = \frac{\text{POSITIONAL TOLERANCE}}{\text{FOR EACH PART.}}$$

Given a situation where an M14×2 bolt is used to fasten two parts together where Part "A" has a clearance hole diameter of 14.4/14.2, and Part "B" is threaded with M14×2 to accommodate the bolt, the positional tolerance is calculated as follows.

$$\frac{\text{MMC HOLE (14.2)} - \text{MMC BOLT (14)}}{2} = \text{POSITIONAL TOLERANCE (0.1)}$$

A drawing representing the positional tolerance calculation for this fixed fastener is shown in Example 8-6.

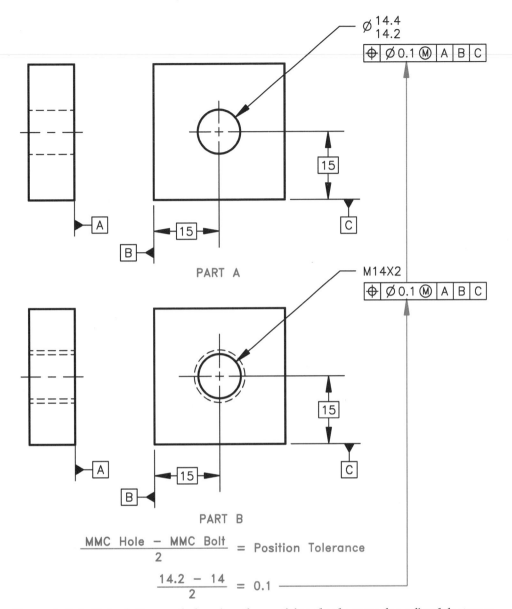

Example 8-6. Calculating and showing the positional tolerance for a fixed fastener.

Sometimes an engineer may want to design the positional tolerance between two or more parts in a fixed fastener situation with a greater amount of positional tolerance applied to the unthreaded part than the threaded part. For example, 70% of the tolerance applied to the unthreaded part and 30% to the threaded part. The revised positional tolerance for Part "A" and Part "B" in Example 8-6 can be calculated using the formula:

MMC HOLE (14.2) – MMC BOLT (14) = 0.2

0.2 × 70% (.70) = 0.14 POSITIONAL TOLERANCE FOR PART A.
0.2 × 30% (.30) = 0.06 POSITIONAL TOLERANCE FOR PART B.

PROJECTED TOLERANCE ZONE

In some situations where positional tolerance is used entirely in out-of-squareness it may be necessary to control perpendicularity and position above the part. The use of a ***projected tolerance zone*** is recommended when variations in perpendicularity of threaded or press-fit holes could cause the fastener to interfere with the mating part. A projected tolerance zone is usually specified for fixed fastener situations, such as the threaded hole for a bolt or the press fit hole of a pin application. The length of a projected tolerance zone may be specified as the distance the fastener extends into the mating part, the thickness of the part, or the height of a press fit stud. The normal positional tolerance extends through the thickness of the part as previously discussed. However, this application may cause an interference between the location of a thread or press-fit object and its mating part. This is because the attitude of a fixed fastener is controlled by the actual angle of the threaded hole. There is no clearance available to provide flexibility. For this reason, the projected tolerance zone is established at true position and extends away from the primary datum at the threaded feature. The projected tolerance zone provides a bigger tolerance because it is projected away from the primary datum, rather than within the thread. The projected tolerance is also easier to inspect than the tolerance applied to the pitch diameter of the thread, because a thread gage with a post projecting above the threaded hole may be used to easily verify the projected tolerance zone with a coordinate measuring machine (CMM). A detailed example of the projected tolerance zone is shown in Example 8-7.

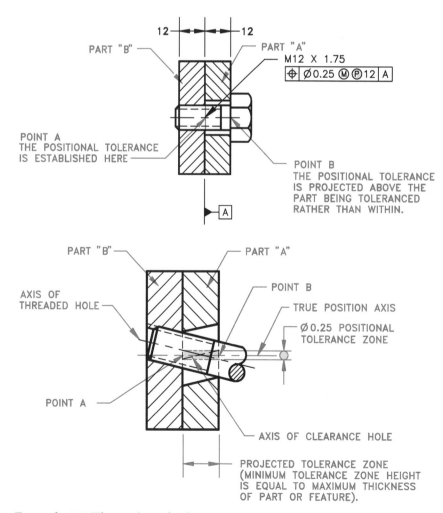

Example 8-7. The projected tolerance zone.

The projected tolerance zone symbol is detailed in Example 8-8. The projected tolerance zone representation may be shown on a drawing a couple of different ways.

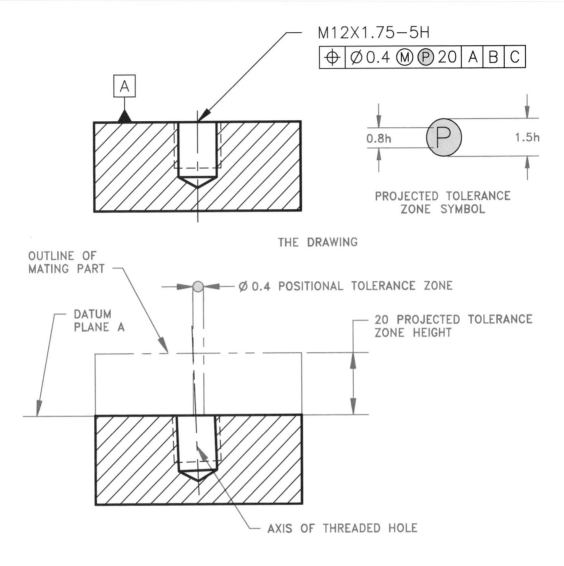

THE DRAWING

THE MEANING

Example 8-8. Projected tolerance zone representation with the length of the projected tolerance zone given in the feature control frame.

One method for displaying the projected tolerance zone is where the projected tolerance zone symbol and height is placed in the feature control frame after the geometric tolerance and related material condition symbol. The related thread specification is then connected to the sectional view of the thread symbol. With this method, the projected tolerance zone is assumed to extend away from the threaded hole at the primary datum. Refer to Example 8-8.

To provide additional clarification, the projected tolerance zone may be shown using a chain line in the view where the related datum appears as an edge and the minimum height of the projection is dimensioned. Refer to Example 8-9. When this is done, the projected tolerance zone symbol is shown alone in the feature control frame after the geometric tolerance and material condition symbol (if any). The meaning is the same as previously discussed.

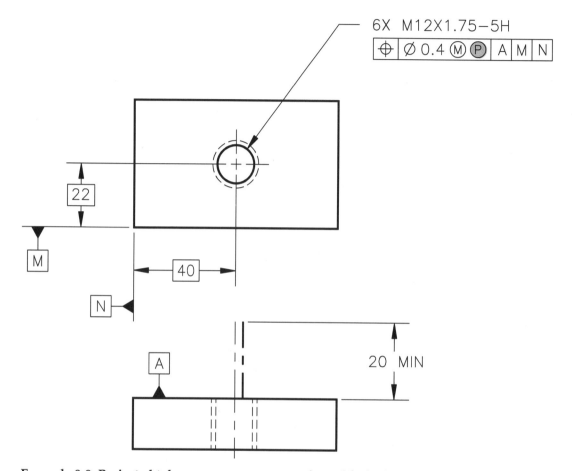

Example 8-9. Projected tolerance zone representation with the length of the projected tolerance zone shown with a chain line and a minimum dimension in the adjacent view.

The projected tolerance zone is often established by a positional tolerance that controls location and perpendicularity. A perpendicularity geometric tolerance may be used to provide a tighter control than that allowed by the positional tolerance, as shown in Example 8-10. Either of the previously discussed representation techniques may be used.

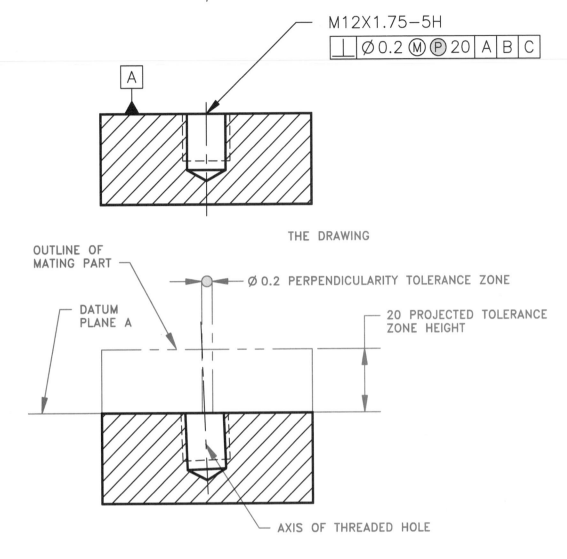

THE DRAWING

THE MEANING

Example 8-10. A perpendicularity geometric tolerance may be used to provide a tighter control than that allowed by the positional tolerance.

VIRTUAL CONDITION

Virtual condition is a boundary that takes into consideration the combined effect of feature size at MMC and geometric tolerance. Virtual condition establishes a *working zone* that is used to establish gage member sizes and the MMC size of mating parts or fasteners for mating parts. The virtual condition represents *extreme conditions* at MMC plus or minus the related geometric tolerance. This is used to determine clearance between mating parts.

It is important to determine the virtual condition when designing mating parts. For example, if a bolt is intended to pass through a hole and the bolt head is to rest flat on the surface, then the bolt diameter can be no bigger than the MMC of the hole less the geometric tolerance. This is the virtual condition. It is not possible to be certain of interchangeability of mating parts if virtual condition is violated. The virtual condition of a feature must be interchangeable with the virtual condition of its mating part. The virtual condition is calculated for situations involving internal or external features.

When calculating the virtual condition of an internal feature, use the formula:

MMC SIZE OF THE FEATURE
– <u>RELATED GEOMETRIC TOLERANCE</u>
= VIRTUAL CONDITION

Given the part shown in Example 8-11, calculate the virtual condition.

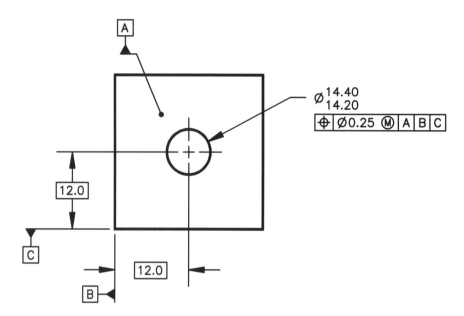

THE DRAWING

THE CALCULATION:

```
                MMC HOLE  =  14.20
   –  GEOMETRIC TOLEANCE  =   0.25
         VIRTUAL CONDITION =  13.95
```

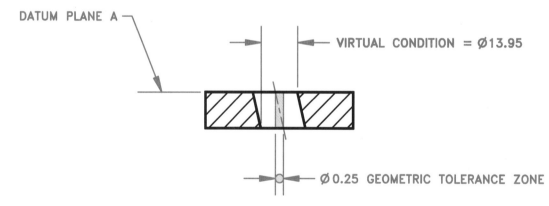

THE MEANING

	POSSIBLE PRODUCED SIZES	GEOMETRIC TOLER-ANCES AT GIVEN PRODUCED SIZES	VIRTUAL CONDITION
MMC	14.20	0.25	13.95
	14.30	0.35	13.95
	14.40	0.45	13.95

Example 8-11. Calculating the virtual condition of an internal feature.

When calculating the virtual condition of an external feature, use the formula:

MMC SIZE OF THE FEATURE
+ RELATED GEOMETRIC TOLERANCE
= VIRTUAL CONDITION

Given the part shown in Example 8-12, calculate the virtual condition.

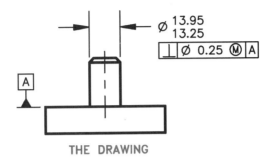

THE DRAWING

THE CALCULATION:

MMC PIN = 13.95
+ GEOMETRIC TOLERANCE = 0.25
VIRTUAL CONDITION = 14.20

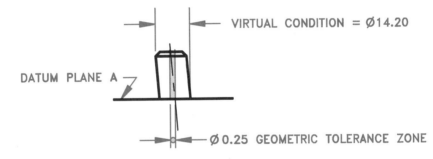

THE MEANING

	POSSIBLE PRODUCED SIZES	GEOMETRIC TOLER- ANCES AT GIVEN PRODUCED SIZES	VIRTUAL CONDITION
MMC	13.95	0.25	14.20
	13.85	0.35	14.20
	13.75	0.45	14.20
	13.65	0.55	14.20
	13.55	0.65	14.20
	13.45	0.75	14.20
	13.35	0.85	14.20
	13.25	0.95	14.20

Example 8-12. Calculating the virtual condition of an external feature.

Zero Positional Tolerance at MMC with the Clearance Hole at Virtual Condition

An application of zero positional tolerance at MMC is used with the design of the maximum material condition of clearance holes at virtual condition, as shown in Example 8-13. Notice that the maximum material condition of the hole is the same as the virtual condition calculated in Example 8-11. As previously explained, virtual condition establishes the working zone that is used to establish the MMC size of mating parts. Therefore, starting with the virtual condition as the maximum material condition of the feature and allowing the produced size to increase with the MMC application is a realistic approach.

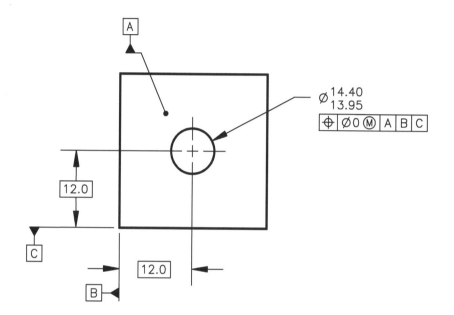

	POSSIBLE PRODUCED SIZES	GEOMETRIC TOLER- ANCES AT GIVEN PRODUCED SIZES	VIRTUAL CONDITION
MMC	13.95	0	13.95
	14.00	0.05	13.95
	14.10	0.15	13.95
	14.20	0.25	13.95
	14.30	0.35	13.95
	14.40	0.45	13.95

Example 8-13. Specifying zero positional tolerance at MMC with the maximum condition of the clearance hole equal to virtual condition.

Virtual Condition of Size Datums

When the axis or center plane of a datum feature of size is controlled by a geometric tolerance. The datum feature implies virtual condition even if a related datum reference is MMC. Refer to Example 8-14 and see that the virtual condition of the feature at Datum D is ⌀24.70. The positional tolerance of the 4×⌀8.0–8.5 holes in reference to Datum D at MMC implies a reference to Datum D at the *virtual* condition and *not* to Datum D at *maximum* condition.

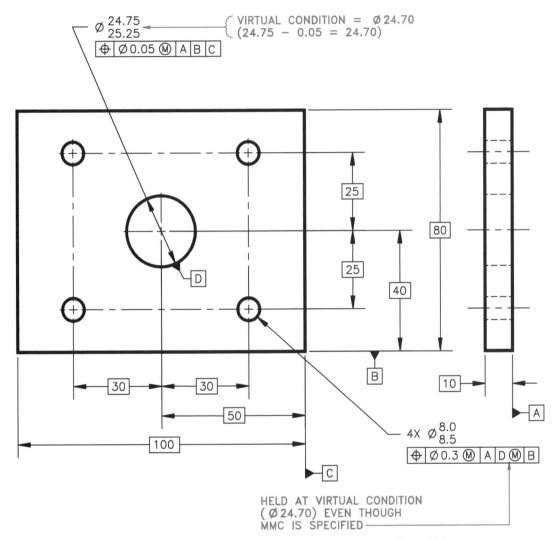

Example 8-14. The application of a datum feature of size at virtual condition.

If you do not want the virtual condition to be applied to the MMC datum reference, then you should consider controlling the datum feature with zero geometric tolerance at MMC. In the case of the drawing in Example 8-14, the positional tolerance associated with Datum D could have been 0 at MMC, in which case the maximum material condition (⌀24.75) of the datum feature would have been equal to the virtual condition.

CONCENTRICITY TOLERANCE

Concentricity is used to establish a relationship between the axes of two or more cylindrical features of an object. The concentricity geometric characteristic symbol is shown detailed in a feature control frame in Example 8-15. Perfect concentricity exists when the axes of each cylindrical feature fall on the same centerline, as shown in Example 8-16.

Example 8-15. The concentricity geometric characteristic symbol in a feature control frame.

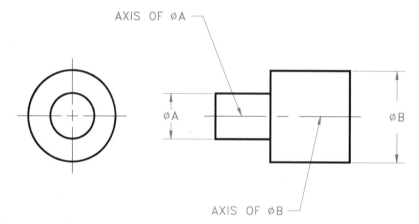

Example 8-16. Perfect concentricity is when the axes of two or more features line up.

Concentricity is the condition where the axis of all cross-sectional elements of a cylindrical surface are common with the axis of a datum feature. The concentricity tolerance specifies a cylindrical (diameter) tolerance zone. The axis of this tolerance zone coincides with a datum axis. All of the median points that originate from the feature surface must be within the cylindrical concentricity tolerance zone, as shown in Example 8-17. The specified concentricity tolerance and related datum reference must apply *only* on an RFS basis.

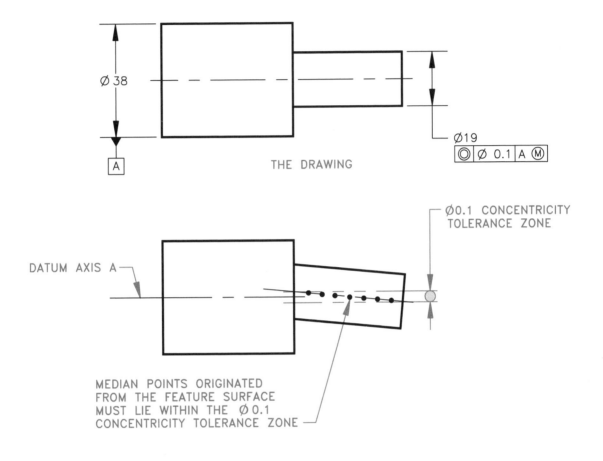

Example 8-17. The concentricity geometric tolerance.

As noted in ASME Y14.5M, irregularities in the form of an actual feature to be inspected may make it difficult to establish the location of a feature's median points. Finding the median points of a feature may require time-consuming analysis of surface variations. It is recommended that runout or positional geometric tolerances be used unless it is absolutely necessary to control a feature's median points.

POSITIONAL TOLERANCING FOR COAXIALITY

Positional tolerancing is recommended over concentricity tolerancing when the control of the axes of cylindrical features can be applied on a material condition basis. A coaxial relationship may be controlled by specifying a positional tolerance at MMC with the datum feature reference specified on an MMC or RFS basis depending on the design requirements. Refer to Example 8-18.

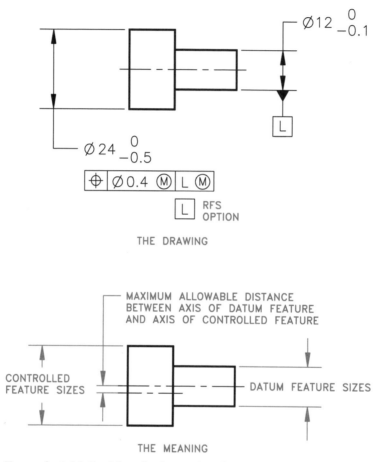

Example 8-18. Positional tolerancing for coaxiality.

When the datum feature is specified on an MMC basis, any departure of the datum feature from MMC may result in an additional displacement of the datum axis and the controlled feature axis. The maximum allowable distance between the axis of the datum feature and the axis of the controlled feature may be calculated at various produced sizes using the following formula:

$$a + b = c$$

Where:

$c =$ Maximum allowable distance between axis of datum feature and axis of controlled feature.

$a = \dfrac{\text{MMC controlled feature} - \text{Actual feature size} + \text{Geometric tolerance}}{2}$

$b = \dfrac{\text{MMC datum feature} - \text{Actual datum feature size}}{2}$

SYMMETRY

The symmetry geometric characteristic symbol is detailed in a feature control frame in Example 8-19.

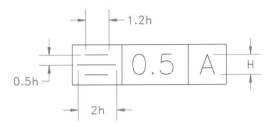

Example 8-19. The symmetry geometric characteristic symbol in a feature control frame.

The symmetry geometric tolerance is a zone where the median points of opposite symmetrical surfaces align with the datum center plane. The symmetry geometric tolerance and related datum reference are applied *only* on an RFS basis. Refer to Example 8-20. The same difficulties discussed in inspecting the median points of a feature for concentricity should also be considered when using symmetry. If this control is not required, then the positional tolerance locating symmetrical features should be considered.

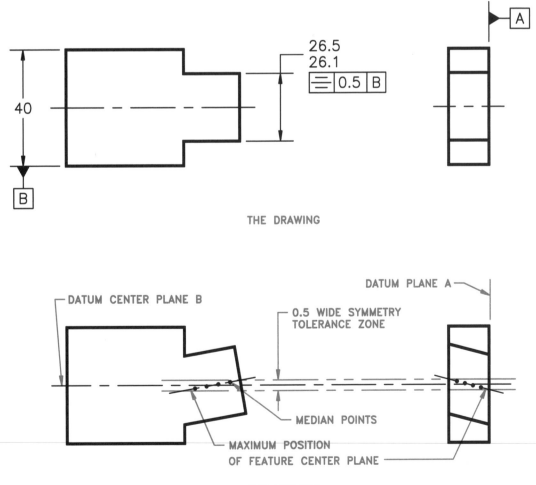

Example 8-20. The symmetry geometric tolerance.

Positional Tolerance Locating Symmetrical Features

Symmetry is a center plane relationship of the features of an object. *Perfect symmetry* or *true position* occurs when the center planes of two or more related symmetrical features line up, as shown in Example 8-21.

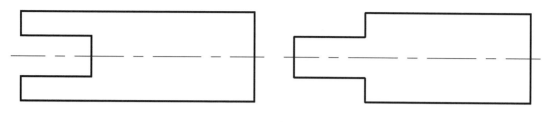

PERFECT SYMMETRY OF A SLOT   PERFECT SYMMETRY OF A TAB

Example 8-21. Perfect positional symmetry is when the center plane of two or more features line up at true position.

A positional tolerance is used when it is required to locate one or more features symmetrically with respect to the center plane of a datum feature, as shown in Example 8-22.

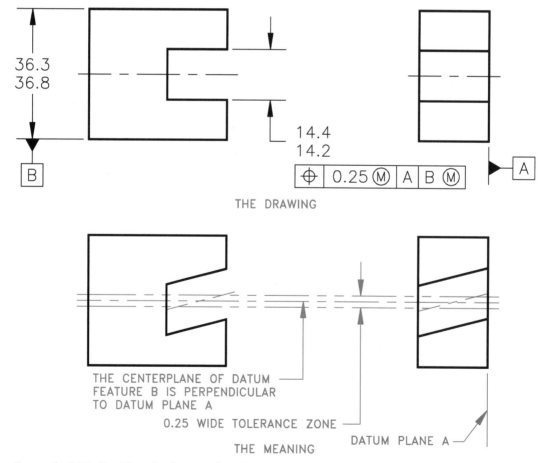

Example 8-22. Positional tolerance locating symmetrical features.

The diameter symbol is omitted in front of the positional tolerance in the feature control frame. This is because the given tolerance zone is the distance between two parallel planes equally divided on each side of true position, rather than a cylindrical tolerance zone, as described in other applications. A material condition symbol MMC or LMC must accompany the positional tolerance, as shown in Example 8-22. Otherwise, RFS is assumed.

Positional tolerancing of symmetrically shaped slots or tabs may be accomplished by identification of related datums, by dimensioning the relationship between slots or tabs, by providing the number of units followed by the size, and by positional tolerance feature control frame. The diameter symbol is omitted from the feature control frame and a material condition symbol is required, as shown in Example 8-23.

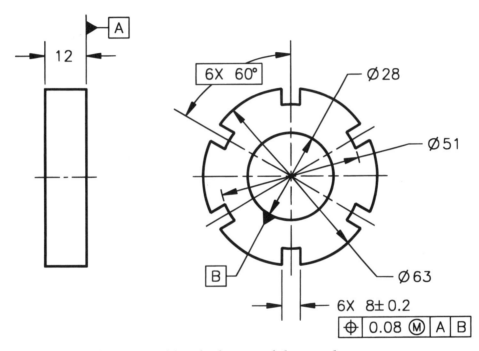

Example 8-23. Showing positional tolerance of slots or tabs.

Zero Positional Tolerance at MMC for Symmetrical Objects

A zero positional tolerance at MMC may be used when it is necessary to control the symmetrical relationship of features within their limits of size. In this application, the datum feature is usually specified on an MMC basis. When the positional controlled feature and the datum feature are at MMC, then perfect symmetry occurs and a boundary of perfect form is established. Out-of-perfect symmetry only happens as the produced size of the features leave MMC. This is similar to the zero positional tolerance at MMC previously discussed.

Name: _____

Test
Location Tolerances
(Part II) and Virtual
Condition

8

1. Symmetry is a _____ relationship of the features of an object.

2. Positional tolerancing of symmetrically shaped slots or tabs may be accomplished by:

    a) Identification of related datums.

    b) Dimensioning the relationship between features.

    c) Providing the number of units followed by the size.

    d) Placing the positional tolerance feature control frame below the size dimension.

    e) All of the above.

3. Give the formula used to determine the positional tolerance of a floating fastener.

    _____

4. Give the formula used to determine the positional tolerance of a fixed fastener.

    _____

    _____

5. Under what condition(s) is a projected tolerance zone recommended? _____

    _____

    _____

    _____

6. Identify the two ways that a projected tolerance zone may be shown on a drawing.

    _____

    _____

    _____

    _____

    _____

    _____

    _____

    _____

    _____

    _____

7. Define "virtual condition." _____

    _____

    _____

8. Give the formula used to calculate the virtual condition of an internal feature.

   _____

   _____

9. Give the formula used to calculate the virtual condition of an external feature.

   _____

   _____

10. _____ may be used to establish a relationship between the axes of two or more cylindrical features of an object.

11. A concentricity tolerance requires the establishment and verification of axes unrelated to surface conditions. Therefore, unless there is a need to control the axis, it is recommended that runout or positional tolerance be used. True or False.

12. A coaxial relationship may be controlled by specifying a positional tolerance at MMC with the datum feature reference specified on an MMC or RFS basis, depending on design requirements. True or False.

13. The application of a geometric tolerance to threaded fasteners is applied to the axis of a cylinder established by the _____ of the thread.

14. What is displayed on the print when the geometric tolerance of a screw thread is applied to the minor diameter? _____

   _____

15. When the axis or center plane of a datum feature of size is controlled by a geometric tolerance the datum feature implies virtual condition even if a related datum reference is MMC. True or False.

16. Concentricity and symmetry must be applied on an RFS basis. True or False.

17. It is recommended that runout or position be used unless it is necessary to control a feature's median points with concentricity. True or False.

18. Describe the symmetry geometric tolerance. _____

   _____

   _____

   _____

19. Calculate the virtual condition of a hole through a part where the hole diameter is ∅14.5±0.3 and the associated positional tolerance is ∅0.1 at MMC. Show your calculations._____

   _____

   _____

20. Calculate the virtual condition of a pin that extends 15mm above the primary datum of a part where the pin diameter is ∅14.5±0.3 and the associated perpendicularity tolerance is ∅0.1 at MMC. Show your calculations. _____

   _____

   _____

Print Reading Exercise

8

Name: _____

The following print reading exercise uses actual industry prints with related questions that require you to read specific dimensioning and geometric tolerancing representations. The answers should be based on previously learned content of this book. The prints used are based on ASME standards, however company standards may differ slightly. When reading these prints, or any other industry prints, a degree of flexibility may be required to determine how individual applications correlate with the ASME standards.

Refer to the print of the BRACKET found on page 304.

1. Refer to the feature control frame associated with the ∅.156±.005 dimension:

    a) Explain the virtual condition of size datum rule that relates to the secondary Datum D at MMC. _____

    _____

    _____

    _____

    _____

    b) Calculate the virtual condition of datum feature D. Show the formula and your calculations. _____

    _____

    _____

    _____

    _____

Refer to the print of the HUB-STATIONARY, ATU found on page 305.

2. Refer to the ∅.352+.005/−.001 dimension:

    a) Why is there no basic dimension symbol around the location dimensions?

    _____

    _____

    _____

    _____

    b) Are the location dimensions placed using rectangular or polar coordinate dimensioning? _____

    _____

    c) What is the positional tolerance? _____

d) Give the positional tolerance at the following actual produced sizes:

Produced Sizes	Positional Tolerance
.357	_____
.355	_____
.353	_____
.351	_____

e) Calculate the virtual condition of these holes. Show the formula and your calculations: _____

_____

_____

3. Refer to the .164-32UNC-2B DEPTH .325 MIN FULL THREAD specification:

a) What is the positional tolerance? _____

b) What does the symbol after MMC in the feature control frame signify?

_____

c) What does the chain line in the sectional view refer to in regard to these features?

_____

_____

d) What does the .325 MIN mean as related to the chain lines discussed in c) above? _____

_____

_____

e) Identify the type of fastener situation if the mating part is an end plate with holes machined to match the six threads in this part, and if a HEX HEAD MACHINE SCREW is used at each of these locations to fasten the parts together.

_____

_____

Refer to the print of the HYDRAULIC VALVE found on page 308.

4. How many different concentricity geometric tolerances are there on this print?

_____

5. Refer to the Ø.961/.959 dimension.

a) Define concentricity as related to the geometric tolerance applied to this dimension. _____

_____

_____

_____

_____

_____

b)   What is the material condition applied to this geometric tolerance? _____

c)   Name the datum reference and give the datum feature dimension. _____

_____

_____

d)   Give the geometric tolerance at the following produced sizes:

Produced Sizes   Geometric Tolerance

.961   _____

.960   _____

.959   _____

Refer to the print of the HOUSING-LENS, FOCUS found on page 313.

6. Refer to the dowel pin associated with the feature on the print identified by general note number 4.

a)   Give the diameter and length of the dowel pin. _____

b)   Give the location of the dowel pin from Datum C. _____

c)   Give the location of the dowel pin from Datum B. _____

d)   What is the positional tolerance for the location of the dowel pin? _____

e)   Assuming the dowel pin size limits are .0628/.0626 based on the American National Standard for dowel pins, calculate the virtual condition. Show the formula and your calculations: _____

_____

_____

_____

Name: _____

Final Exam

Geometric Dimensioning and Tolerancing

PART I

Below is a list of short descriptions with a list of words and symbols at the right. Place the letter of the word or symbol that matches the description in the blank. Each letter may be used more than once. Some selections may not be used.

__U__ 1. Used to define a zone where the center axis or center plane of a feature of size is permitted to vary from true position.

__S__ 2. The general term applied to a physical portion of a part.

__J__ 3. Condition where a feature of size contains the maximum amount of material within the limits.

__Q__ 4. Datum feature symbol.

__L__ 5. Considered a theoretically perfect dimension.

__S__ 6. The actual surface of an object that is used to establish a datum plane.

__R__ 7. All datum planes on a part intersecting at right angles are 90° basic by interpretation.

__C__ 8. A geometric tolerance or datum reference applies at any increment of size of the feature within its size tolerance.

__M__ 9. This geometric tolerance specifies a zone where the required surface element or axis must lie.

__A__ 10. A geometric characteristic with a tolerance zone between two parallel planes and perpendicular to a datum.

__I__ 11. A geometric characteristic used to identify a location tolerance.

__D__ 12. A geometric characteristic with a tolerance zone between two parallel planes and parallel to a datum.

__E__ 13. A single element form control that establishes a tolerance zone between two concentric circles.

__F__ 14. A profile tolerance that is split equally on each side of the true profile.

__K__ 15. A geometric characteristic that establishes two perfectly concentric cylinders where the actual surface must lie.

(A) $\perp$

(B) Datum Target

(C) RFS

(D) //

(E) ◯

(F) Bilateral

(G) ——

(H) LMC

(I) ⊕

(J) MMC

(K) ⌀

(L) Basic

(M) ⌒

(N) Unilateral

(O) ↗

(P) Datum Feature

(Q) [A] ▲

(R) Three Plane Concept

(S) Feature

(T) Positional Tolerance

(U) Simulated Datum

(V) Virtual Condition

PART II

16. The symbol below is called ___DATUM FEATURE SYMBOL___. Fill in the dimensions below as related to lettering height = h when this symbol is properly placed on a drawing.

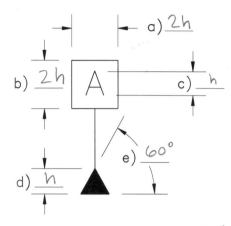

a) 2h
b) 2h
c) h
d) h
e) 60°

17. The symbol below is called ___FEATURE CONTROL FRAME___. Fill in the dimensions below as related to lettering height = h when this symbol is properly placed on a drawing.

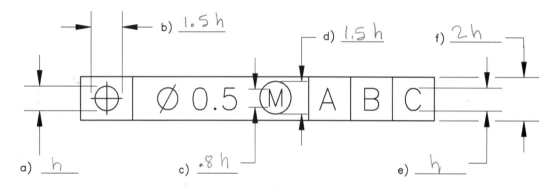

a) h
b) 1.5 h
c) .8 h
d) 1.5 h
e) h
f) 2h

18. Identify the items in the symbol below and write your answers in the blanks on the right.

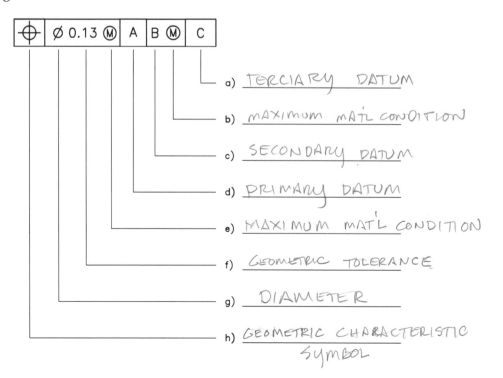

a) TERCIARY DATUM
b) MAXIMUM MAT'L CONDITION
c) SECONDARY DATUM
d) PRIMARY DATUM
e) MAXIMUM MAT'L CONDITION
f) GEOMETRIC TOLERANCE
g) DIAMETER
h) GEOMETRIC CHARACTERISTIC SYMBOL

19. Name each of the dimensioning symbols shown below. Place your answer on the blank provided to the right of the symbol.

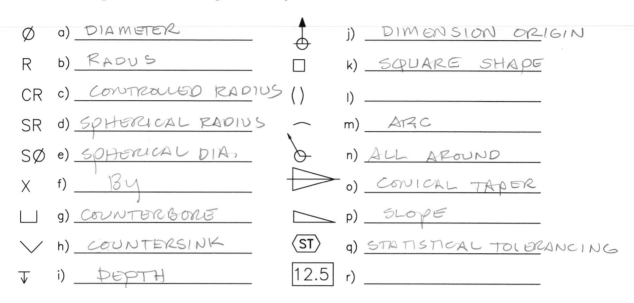

Ø a) DIAMETER

R b) RADUS

CR c) CONTROLLED RADIUS

SR d) SPHERICAL RADIUS

SØ e) SPHERICAL DIA,

X f) By

⊔ g) COUNTERBORE

∨ h) COUNTERSINK

↧ i) DEPTH

j) DIMENSION ORIGIN

k) SQUARE SHAPE

( ) l) _____

⌒ m) ARC

n) ALL AROUND

o) CONICAL TAPER

p) SLOPE

⟨ST⟩ q) STATISTICAL TOLERANCING

12.5 r) _____

20. Name the following symbols for the geometric characteristics. Place your answer on the blank to the right of each symbol.

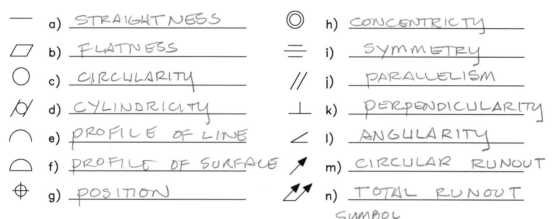

— a) STRAIGHTNESS

▱ b) FLATNESS

○ c) CIRCULARITY

⌀ d) CYLINDRICITY

⌒ e) PROFILE OF LINE

◠ f) PROFILE OF SURFACE

⊕ g) POSITION

◎ h) CONCENTRICTY

≡ i) SYMMETRY

// j) PARALLELISM

⊥ k) PERPENDICULARITY

∠ l) ANGULARITY

↗ m) CIRCULAR RUNOUT

↗↗ n) TOTAL RUNOUT

21. The symbol below is called DATUM TARGET SYMBOL. Identify the symbol components in the blanks provided and provide the dimension related to lettering height = h when this symbol is properly placed on a drawing.

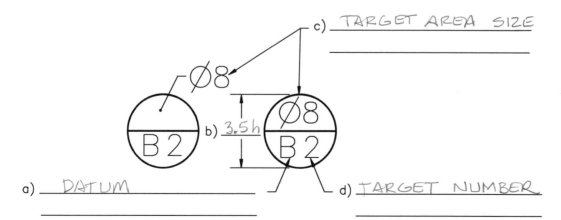

c) TARGET AREA SIZE

_____

b) 3.5 h

a) DATUM

_____

d) TARGET NUMBER

22. List three items that may be identified as datums.

    SURFACE

    POINT

    AXIS

23. Draw or neatly sketch the symbol for each of the items listed below.

    Maximum Material Condition ⓂM

    Projected Tolerance Zone ⓅP

    Least Material Condition ⓁL

24. Name the following datum related symbols. Place your answer in the blanks provided below the symbol.

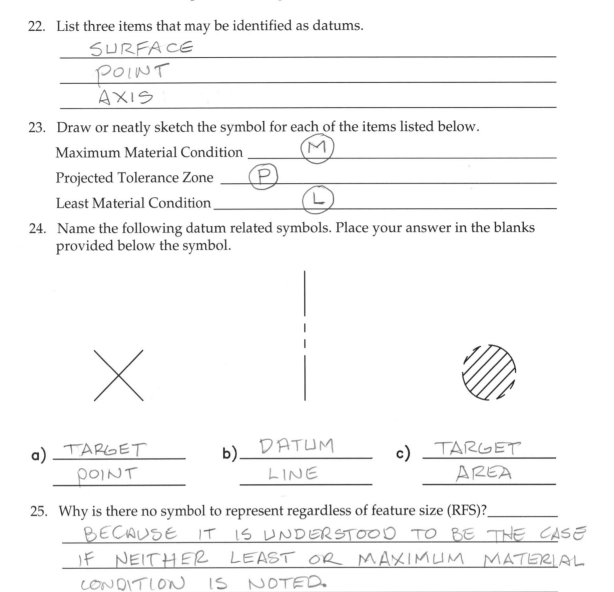

    a) TARGET POINT   b) DATUM LINE   c) TARGET AREA

25. Why is there no symbol to represent regardless of feature size (RFS)?

    BECAUSE IT IS UNDERSTOOD TO BE THE CASE IF NEITHER LEAST OR MAXIMUM MATERIAL CONDITION IS NOTED.

PART III

26. Given the object shown below, answer the following questions.

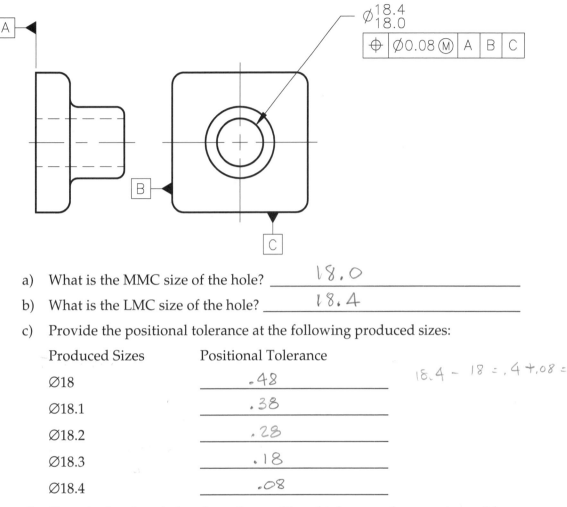

a) What is the MMC size of the hole? ____18.0____

b) What is the LMC size of the hole? ____18.4____

c) Provide the positional tolerance at the following produced sizes:

Produced Sizes	Positional Tolerance	
Ø18	.48	18.4 − 18 = .4 +.08 =
Ø18.1	.38	
Ø18.2	.28	
Ø18.3	.18	
Ø18.4	.08	

d) Show in the chart below how the positional tolerance changes at possible produced sizes if the following feature control frames are substituted:

a) ⊕ | Ø0.08 | A | B | C

b) ⊕ | Ø0.08 ⓁĹ | A | B | C

c) ⊕ | Ø0 Ⓜ | A | B | C

Produced Sizes	Positional Tolerance		
	a) RFS	b) LMC	c) 0 at MMC
Ø18.0	_____	_____	_____
Ø18.1	_____	_____	_____
Ø18.2	_____	_____	_____
Ø18.3	_____	_____	_____
Ø18.4	_____	_____	_____

27. Given the object shown below, answer the following questions:

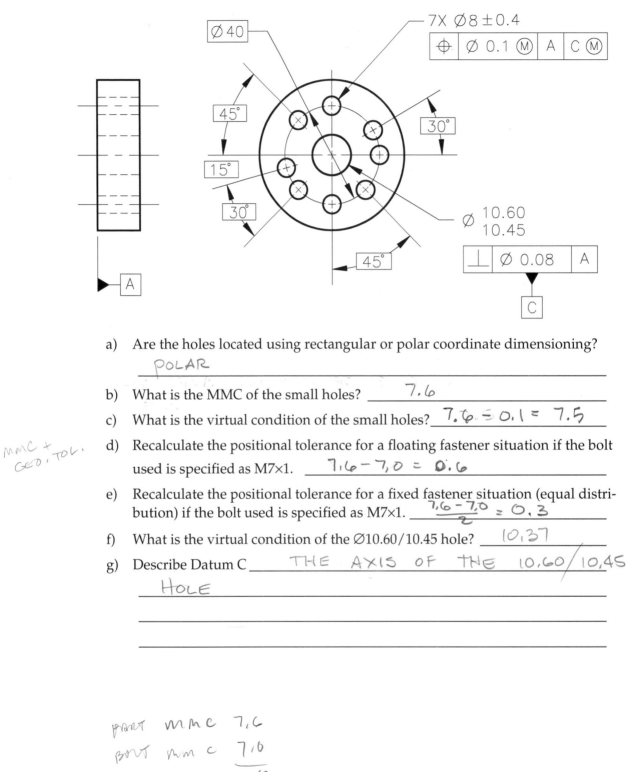

a) Are the holes located using rectangular or polar coordinate dimensioning?

POLAR

b) What is the MMC of the small holes? _____ 7.6 _____

c) What is the virtual condition of the small holes? 7.6 ÷ 0.1 = 7.5

MMC +
GEO. TOL.

d) Recalculate the positional tolerance for a floating fastener situation if the bolt used is specified as M7×1. _____ 7.6 − 7.0 = 0.6 _____

e) Recalculate the positional tolerance for a fixed fastener situation (equal distribution) if the bolt used is specified as M7×1. $\frac{7.6 - 7.0}{2} = 0.3$

f) What is the virtual condition of the Ø10.60/10.45 hole? _____ 10.37 _____

g) Describe Datum C _____ THE AXIS OF THE 10.60/10.45

HOLE

PART MMC 7.6
BOLT MMC 7.0
    .6

28. Provide a short, complete interpretation of the feature control frame associated with the following drawing. __PROFILE OF LINE BETWEEN POINTS X & Y WITH A BILATERAL TOLERANCE ZONE OF 0.2 WITH RESPECT TO DATUM A.__

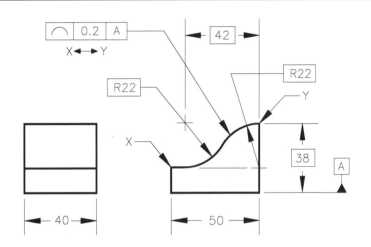

29. Read the following drawing and give a short, complete description of the geometric tolerance. __PROFILE OF THE SURFACE ALL AROUND THE PART WITH A BILATERAL TOLERANCE ZONE OF 0.4 WITH RESPECT TO DATUM A.__

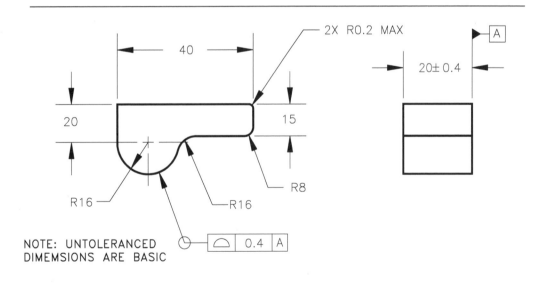

NOTE: UNTOLERANCED DIMEMSIONS ARE BASIC

30. Read the following drawing and give a short, complete description of the geometric tolerance. PROFILE OF SURFACE BETWEEN POINTS X & Y WITH A UNILATERAL OUTSIDE TOLERANCE ZONE OF 0.4 WITH RESPECT TO DATUM A

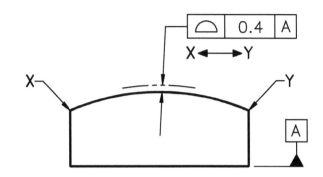

31. Read the following drawing and give a short, complete description of the geometric tolerance. TOTAL RUNOUT OF THE 50 LONG SURFACE LOCATED WHERE THE CHAIN LINE IS POSITIONED WITH A BILATERAL TOLERANCE OF 0.08 WITH RESPECT TO DATUM AXIS G-H

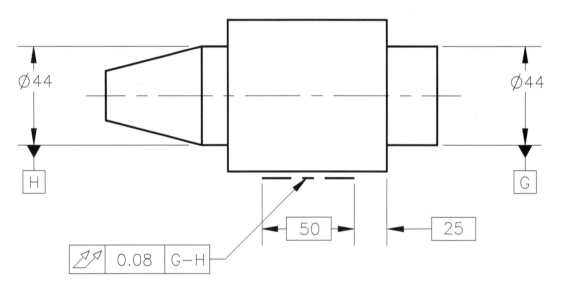

32. Read the following drawing and give a short, complete description of the geometric tolerance. ① PROFILE OF LINE BETWEEN POINT X & Y WITH A UNILATERAL OUTSIDE TOLERANCE ZONE OF 0.2

② CIRCULAR RUNOUT OF SURFACE WITH A BILATERAL TOLERANCE OF 0.08 WITH RESPECT TO DATUMS A & B

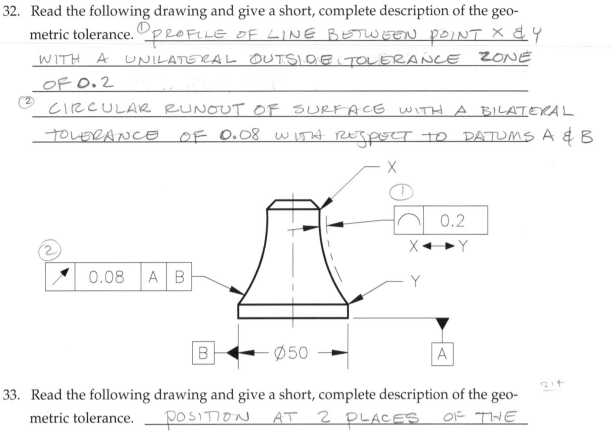

33. Read the following drawing and give a short, complete description of the geometric tolerance. POSITION AT 2 PLACES OF THE Ø16 HOLES WITH A TOLERANCE FOR THE AXIS TO BE WITHIN A CYLINDRICAL ZONE OF Ø0.3 AT THE MAXIMUM MATERIAL CONDITION WITH RESPECT TO DATUMS A & B (GROUP PATTERN) AND: POSITION OF THE INDIVIDUAL HOLES WITH RESPECT TO EACH OTHER WITH A CYLINDRICAL ZONE OF Ø0.1 AT THE MMC

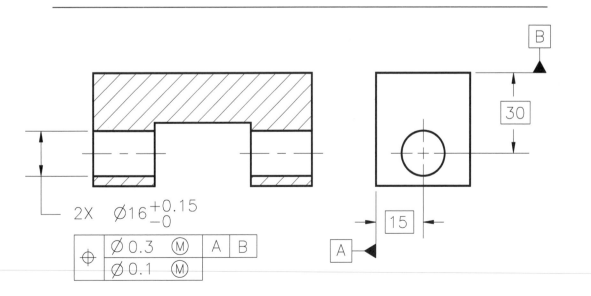

34. Read the following drawing and give a short, complete description of the geometric tolerance. _____

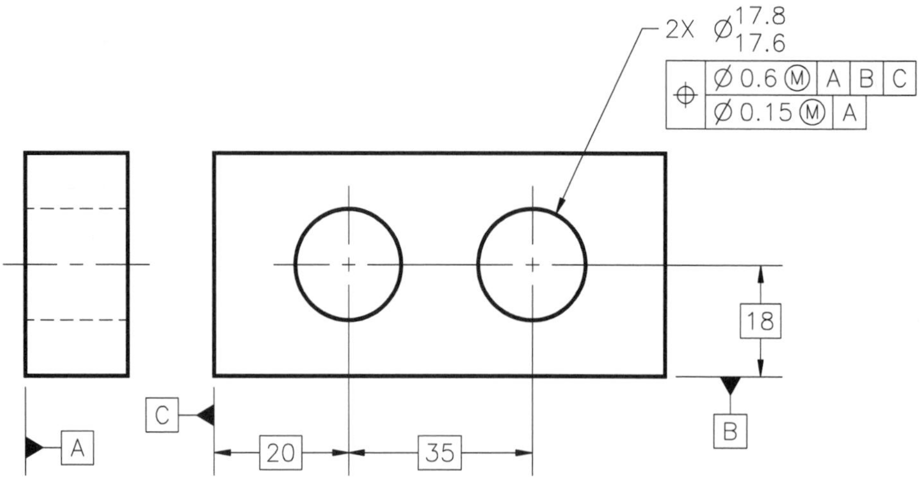

_____

_____

_____

_____

_____

_____

_____

_____

_____

_____

_____

_____

_____

_____

35. Read the following drawing and give a short, complete description of the geometric tolerance. _____

_____

_____

_____

_____

_____

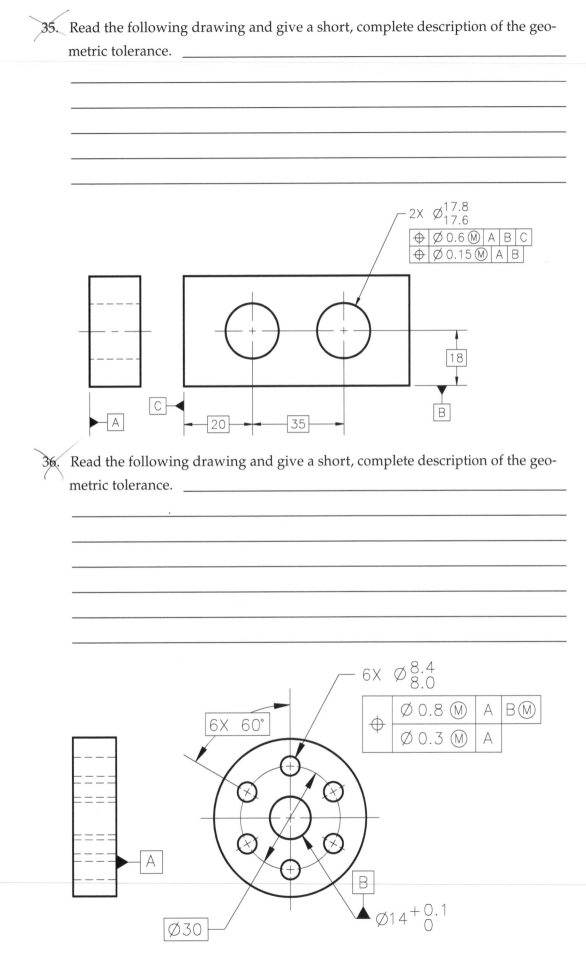

36. Read the following drawing and give a short, complete description of the geometric tolerance. _____

_____

_____

_____

_____

_____

_____

37. Read the following drawing and give a short, complete description of the geometric tolerance. _____

_____

_____

_____

_____

_____

_____

_____

_____

_____

_____

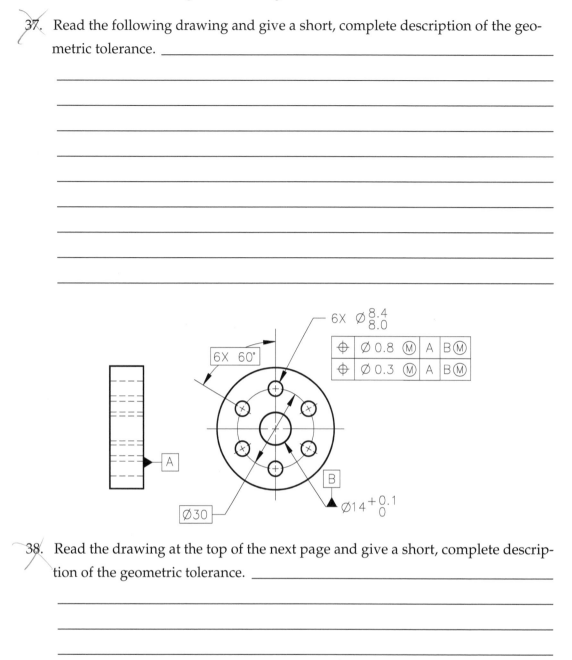

38. Read the drawing at the top of the next page and give a short, complete description of the geometric tolerance. _____

_____

_____

_____

_____

_____

_____

_____

_____

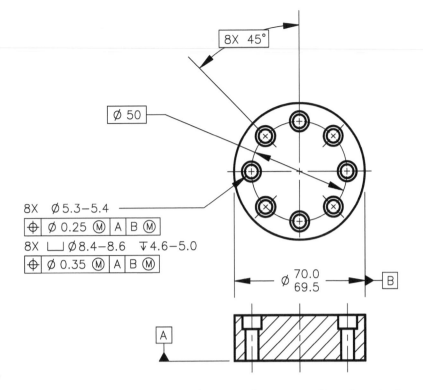

8X 45°

Ø 50

8X Ø 5.3–5.4

| ⌖ | Ø 0.25 Ⓜ | A | B Ⓜ |

8X ⌴ Ø 8.4–8.6 ↧ 4.6–5.0

| ⌖ | Ø 0.35 Ⓜ | A | B Ⓜ |

Ø 70.0 / 69.5   B

A

39. Read the following drawing and give a short, complete description of the geometric tolerance. _____

_____

_____

_____

_____

M12X1.75–5H

| ⌖ | Ø 0.4 Ⓜ Ⓟ | A | M | N |

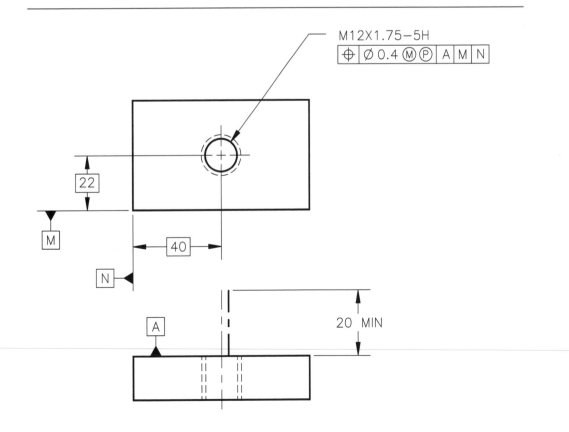

22

M

40

N

A

20 MIN

PART IV

Circle T if the statement is true. Circle F if the statement is false.

T F 40. Unit straightness may be used if the part must be controlled per unit of measure as well as over the total length.

T F 41. Specific area flatness should be avoided on very large parts.

T F 42. The zone descriptor of a circularity tolerance is diameter.

T F 43. Cylindricity is identified by a radius tolerance zone that establishes two perfect concentric cylinders.

T F 44. The profile of a line must be established between two given points on an object.   p 140

T F 45. A parallelism tolerance zone must be between the size tolerance of the feature.   p 158

T F 46. The perpendicularity of a shaft, such as a stud or pin, to a datum feature establishes a cylindrical diameter tolerance zone.

T F 47. The term "EACH ELEMENT" must be applied to a perpendicularity feature control frame.

T F 48. An angularity tolerance must have a basic angular relationship to a datum.

T F 49. The tolerance zone descriptor of a concentricity tolerance is R.   243

T F 50. Concentricity is used to establish the relationship between the axis of two or more cylindrical features of an object. The median points that originate from the feature's surface must be within a cylindrical tolerance zone.

T F 51. The profile tolerance zone may be bilateral or unilateral.

T F 52. Surface straightness may violate perfect form at MMC.

T F 53. Geometric tolerances and related references imply RFS unless otherwise specified.

T F 54. A concentricity tolerance should be used if there is a need to control the axis as in a dynamically balanced shaft, otherwise it is recommended that a runout or positional tolerance be used.

T F 55. A coaxial relationship may be controlled by a positional tolerance at MMC with the datum reference at MMC or RFS.

T F 56. LMC is often used to control minimum edge distance.

T F 57. True position is the theoretically exact location of a feature.

T F 58. The datum reference frame exists in theory. The theoretical reference frame is simulated by positioning the part on datum features to adequately relate the part to the datum reference frame and to restrict motion of the part relative to the reference frame.

T F 59. When reference is made to the datum reference frame, the primary datum should be given first, followed by the secondary and tertiary. This is referred to as datum precedence.

PART V

Given the following specifications, calculate the required unknown values.

60. Given: A shaft with a diameter of 36±0.2.
    Calculate:

    a) Tolerance ___0.4___

    b) MMC ___36.2___

    c) LMC ___35.8___

61. Given: A hole with a diameter of $36.5^{+0.3}_{-0.1}$
    Calculate:

    a) Tolerance ___0.4___

    b) MMC ___36.4___

    c) LMC ___36.8___

    d) Allowance with the shaft in number 60. Show the formula(s) and your
       calculations. ___36.4 - 36.2 = 0.2___

    e) Clearance with the shaft in number 60. Show the formula(s) and your
       calculations. ___36.8 − 35.8 = 1.0___
       ___LMC HOLE − LMC SHAFT = CLEARANCE___

62. Given: Two parts to be bolted together with a 5±0.4 hole through each part and
    an M4.5×0.75 HEX SOCKET HEAD CAP SCREW and HEX NUT for fastening.
    Show the formula(s) and calculations for calculating the following:

    a) Positional tolerance. ___MMC HOLE = 4.6 ; MMC SCREW = 4.5___
       ___4.6 − 4.5 = 0.1___

    b) Virtual condition. ___MMC − GEOM. TOL___
       ___4.6 − 0.0 = 4.6___

63. Given: Two parts to be bolted together. One part has a hole with a diameter of 16.5–16.1. The other part has a threaded hole, M16×2, located to align with the hole through the first part. An M16×2 STANDARD METRIC HEAVY HEX SCREW is used to fasten the two parts together.
    Calculate the positional tolerance for each part based on the following:

    a)   Equally distributed positional tolerance.

    Show the formula(s) and your calculations. _____

    $$\frac{\text{MMC HOLE } (16.1) - \text{MMC BOLT } (16)}{2} = 0.05$$

    b)   Provide 60% of the positional tolerance to the threaded part. Determine position tolerance for each part. Show the formula(s) and your calculations.

    MMC HOLE (16.1) – MMC BOLT (16) = 0.1
    0.1 × .6   THRD PART   = 0.06
    0.1 × .4   NON-THRD PART = 0.04

64. Given: A pin with a diameter of $8 \, {}^{0}_{-0.5}$ is held perpendicular to datum surface A by Ø0.2.
    Calculate:

    a)   MMC pin. ____8.0____

    b)   LMC pin. ____7.5____

    c)   Virtual condition. Show the formula(s) and your calculations.

    MMC + GEOM. TOL
    8.0 + 0.2 = 8.2

Drafting Problems

Geometric Dimensioning and Tolerancing

GENERAL INFORMATION AND INSTRUCTIONS

A variety of drafting problems are provided on the following pages. Drafting problems vary in complexity and require different GD&T applications. The drafting problems are presented as 3-D illustrations, 2-D layouts, or rough engineer's sketches. The GD&T applications are displayed on the problem or in written instructions. Use the following instructions unless otherwise specified with the problem:

- ❏ Use manual or computer-aided design and drafting as specified by your course requirements.
- ❏ Select a drawing scale that clearly displays the features and dimensions.
- ❏ Select drafting sheet sizes that avoid crowding and are in accordance with standard practices such as scale, number of views, amount of dimensions and notes, and space for future revisions.
- ❏ Prepare formal drawings using properly selected multiviews (orthographic projection). The number of views needed depends on the requirements of each drafting problem. This is to be determined by you.
- ❏ Use proper sectioning techniques as needed.
- ❏ Place conventional dimensioning and geometric dimensioning and tolerancing as specified in ASME Y14.5M and as instructed in this workbook.
- ❏ Use unidirectional dimensioning unless otherwise specified by your instructor or other specific instructions.
- ❏ Do metric drawings in millimeters and inch drawings in inches, unless directed by your instructor to convert millimeters to inches or inches to millimeters.
- ❏ Place the following general notes in the lower left corner (1/2″ each way from the corner) unless otherwise specified by your instructor:

4. OTHER NOTES AS NEEDED FOR PROBLEM REQUIREMENTS.

3. REMOVE ALL BURRS AND SHARP EDGES.

2. UNLESS OTHERWISE SPECIFIED, ALL DIMENSIONS ARE IN MILLIMETERS. (or INCHES.)

1. INTERPRET DIMENSIONS AND TOLERANCES PER ASME Y14.5M—1994.

NOTES

❏ Provide specified dimensions as given. Provide general title block tolerances for unspecified dimensions as follows, unless otherwise specified by your instructor:

Inches

$$.X = \pm.1$$
$$.XX = \pm.01$$
$$.XXX = \pm.005$$
$$ANGULAR = \pm30'$$
$$FRACTIONAL = \pm1/32$$
$$FINISH = 125\mu IN$$

Metric

$$.X = \pm0.5$$
$$.XX = \pm0.2$$
$$.XXX = \pm0.1$$
$$ANGULAR = \pm30'$$
$$FINISH = 8\mu M$$

❏ Use line standards as recommended in ASME Y14.2M.

DRAFTING PROBLEM 1

Units
METRIC

Application
Entry level, datum feature symbols, basic dimensions, feature control frame.

Name
FACE BLOCK

Material
MILD STEEL (MS)

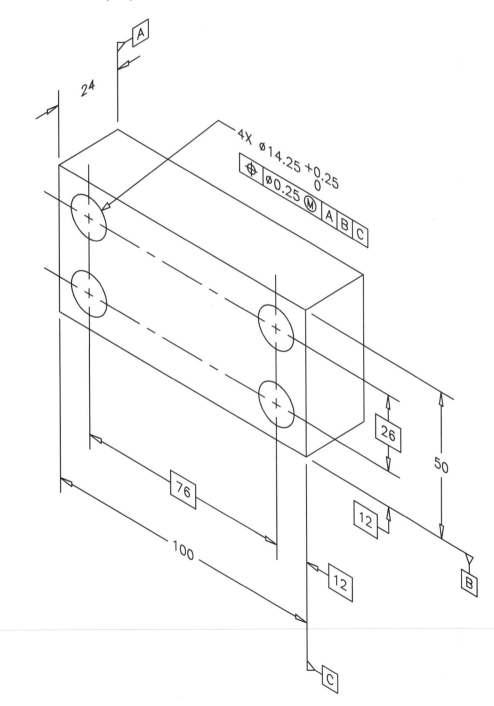

DRAFTING PROBLEM 2

Units
INCH

Application
Entry level, datum feature symbols, basic dimension, feature control frames. The words "FRONT SURFACE" and "BACK SURFACE" should not be placed on the drawing.

Name
GAUGE BLOCK

Material
SAE 4320, .315 IN. THICK

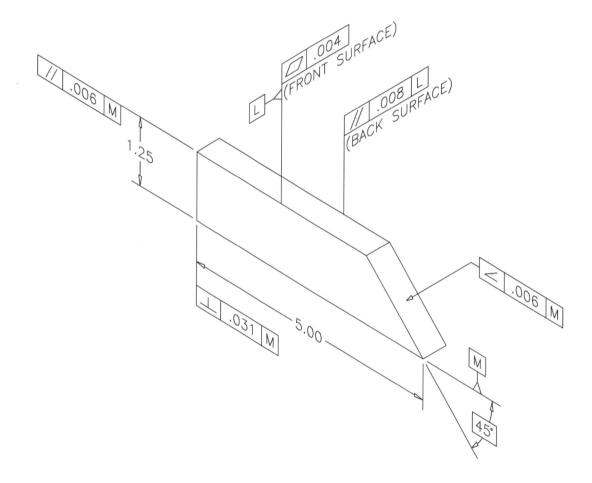

DRAFTING PROBLEM 3

Units
METRIC

Application
Entry level, circularity and cylindricity. A section is shown here only for clarity.

Name
VALVE PIN

Material
PHOSPHOR BRONZE

Finish
ALL OVER 0.2µM

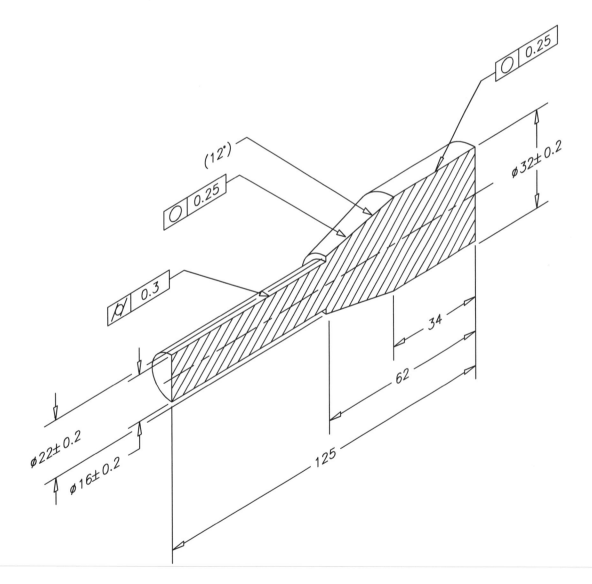

DRAFTING PROBLEM 4

Units
   METRIC

Application
   Entry level, unilateral profile of a surface, parallelism.

Name
   INSERT

Material
   SAE 4640

Finish
   ALL OVER 0.2μM

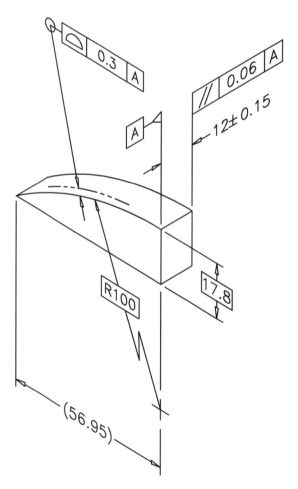

DRAFTING PROBLEM 5

Units
METRIC

Application
Entry level, profile between points, parallelism.

Name
SPRING CLIP

Material
SAE 1060, 10mm THICK

Finish
ALL OVER 0.5μM

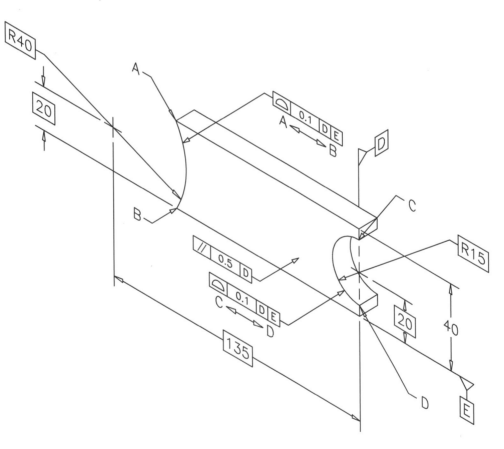

DRAFTING PROBLEM 6

Units
INCH

Application
Entry level, profile of an angled surface. A section is shown here for clarity.

Name
POSITION PIN

Material
SAE 2330

Finish
ALL OVER 32μIN

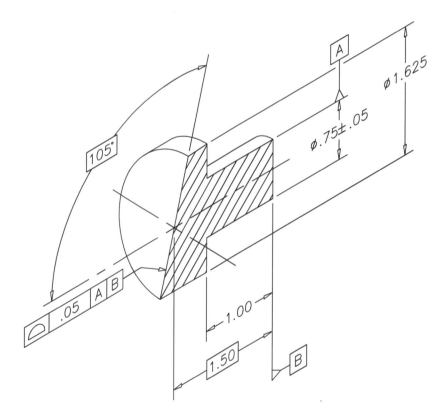

DRAFTING PROBLEM 7

Units
INCH

Application
Medium level, profile coplanar surfaces, perpendicularity, position, center plane datum. The object in the problem is displayed in full section for clarity.

Name
FLUSH PLATE

Material
SAE 1137

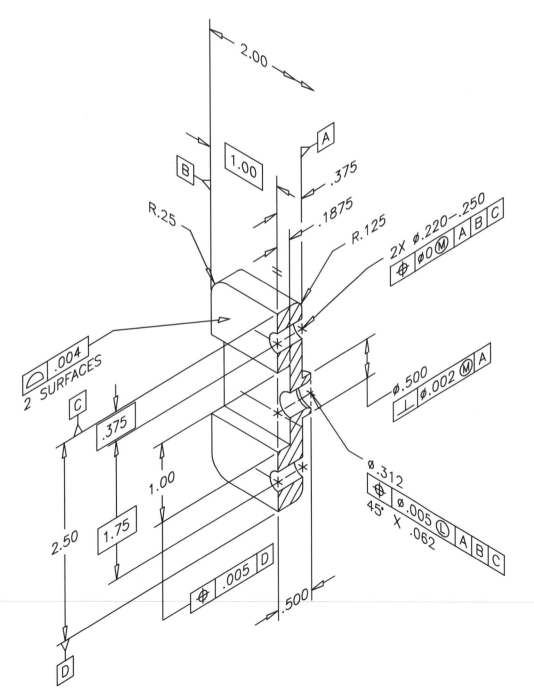

DRAFTING PROBLEM 8

Units
   INCH

Application
   Medium level, inclined datum features, angularity, perpendicularity, position.

Name
   STEERING ARM STOP

Material
   SAE 4815

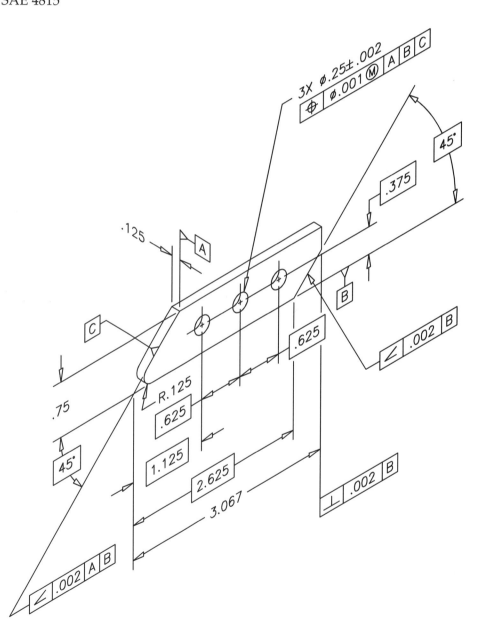

DRAFTING PROBLEM 9

Units
INCH

Application
Medium level, axis parallelism. The object in the problem is displayed in full section for clarity.

Name
SHAFT BEAM

Special Instructions
Provide R12 fillets.

Material
SAE 1070

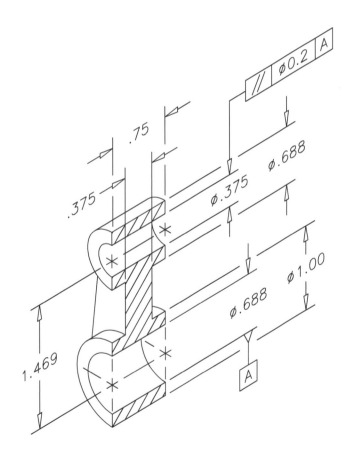

DRAFTING PROBLEM 10

Units
INCH

Application
Medium level, circular runout to two datum diameters. The object in the problem is displayed in full section for clarity.

Name
VALVE PIN

Material

BRONZE

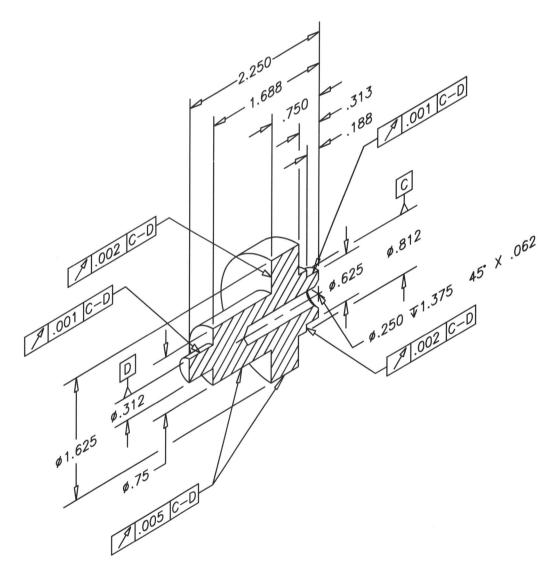

DRAFTING PROBLEM 11

Units
METRIC

Application
Medium level, runout relative to a datum surface and diameter with form control specified.

Name
COVER

Material
SAE 1015

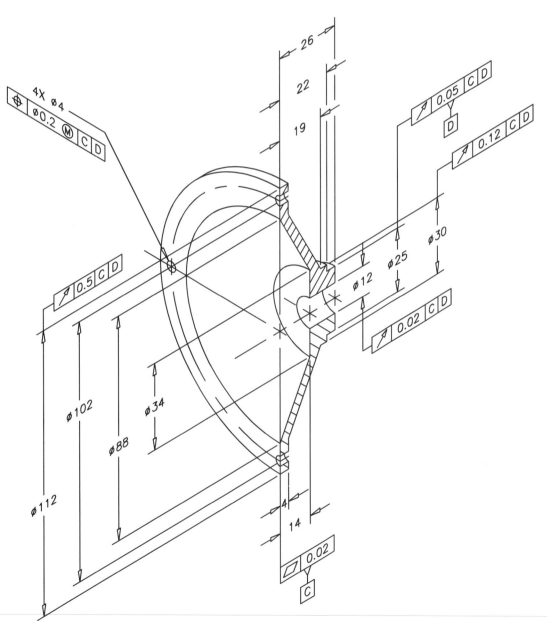

DRAFTING PROBLEM 12

Units
  INCH

Application
  Medium level, at LMC.

Name
  THRUST WASHER

Material
  SAE 5150

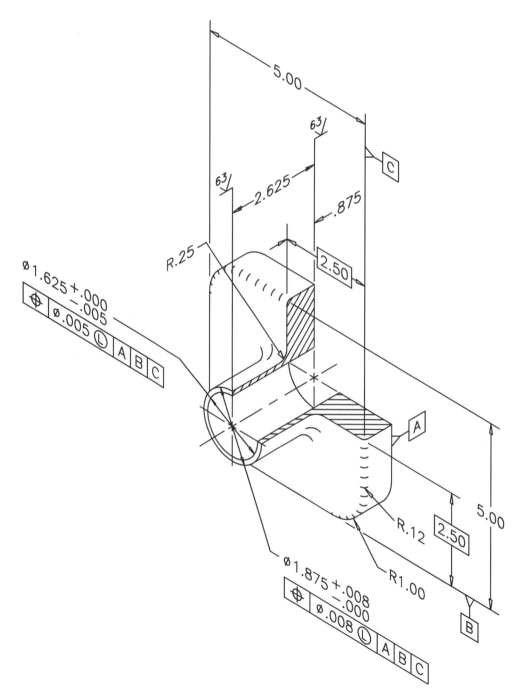

DRAFTING PROBLEM 13

Units
METRIC

Application
Entry level, zero positional tolerance at MMC for pattern of holes.

Name
PLATE

Material
SAE 1020

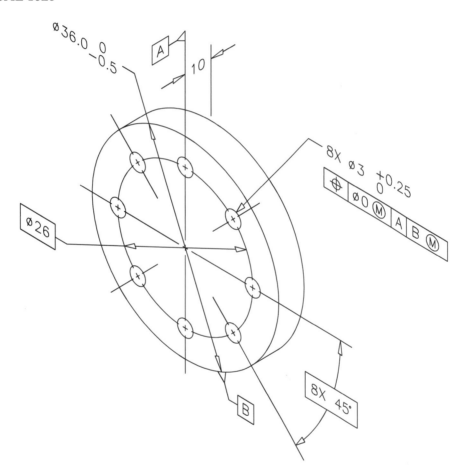

DRAFTING PROBLEM 14

Units
METRIC

Application
Medium level, position holes and slots, partial written instructions.

Name
MODULAR CHASSIS PLATE

Material
SAE 30308, 1.5 THICK

Finish
ALL OVER 0.2μM

Additional Instructions
1. Provide the note "UNTOLERANCED DIMENSIONS LOCATING TRUE POSI-TION ARE BASIC" on the drawing.
2. The 6×∅8 and the 4×∅5 holes shall have positional tolerance of 0.25 at MMC to datums A, B, and C.
3. The slots (12×6 with full radius ends) shall have a positional tolerance of 0.5 at MMC applied to the 12 dimension, and a positional tolerance of 0.25 at MMC applied to the 6 dimension. Note the "BOUNDARY" reference.

Optional Instructions
This drawing may be done using arrowless dimensioning depending on your course objectives and instructions.

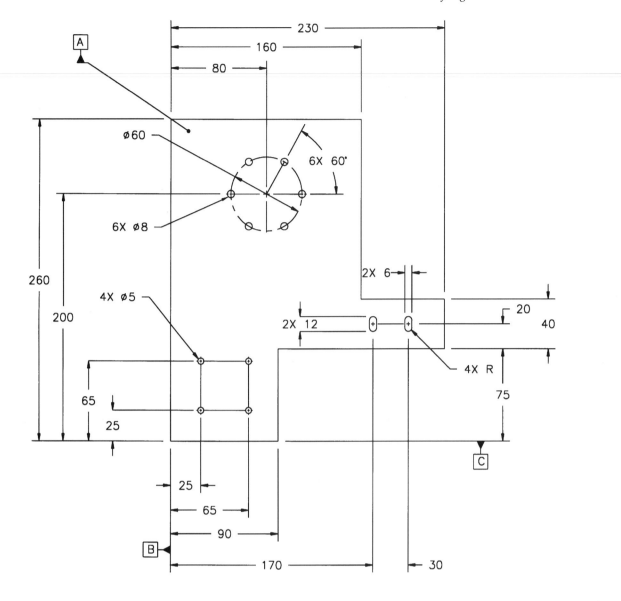

DRAFTING PROBLEM 15

Units
METRIC

Application
Medium level, positional tolerance for symmetry, positional tolerance, perpendicularity.

Name
STOP PIN

Material
SAE 3240

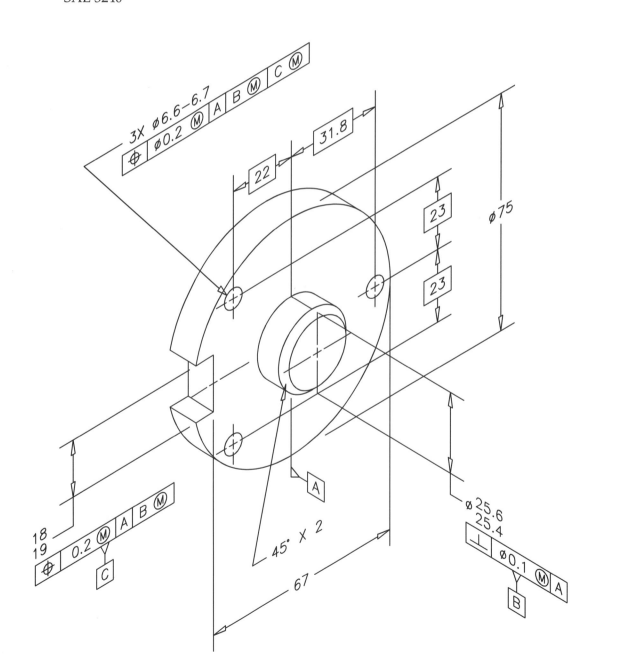

DRAFTING PROBLEM 16

Units
METRIC

Application
Medium level, position holes, form, concentricity, orientation.

Name
HUB INSERT

Material
SAE 1045

Finish
ALL OVER 1.6μM

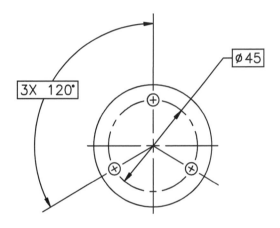

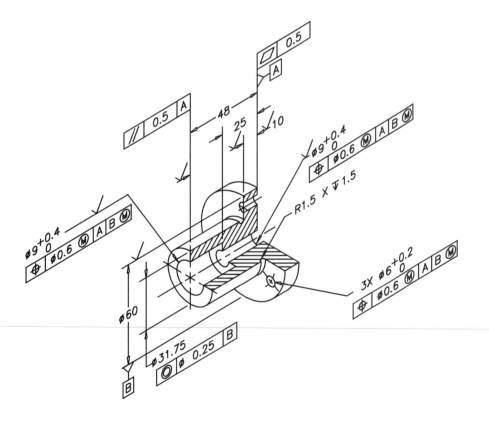

DRAFTING PROBLEM 17

Units
INCH

Application
Advanced level, partial surface datum, positional tolerance repetitive features, datum target areas and line.

Name
EXTENSION SUPPORT

Material
ALUMINUM

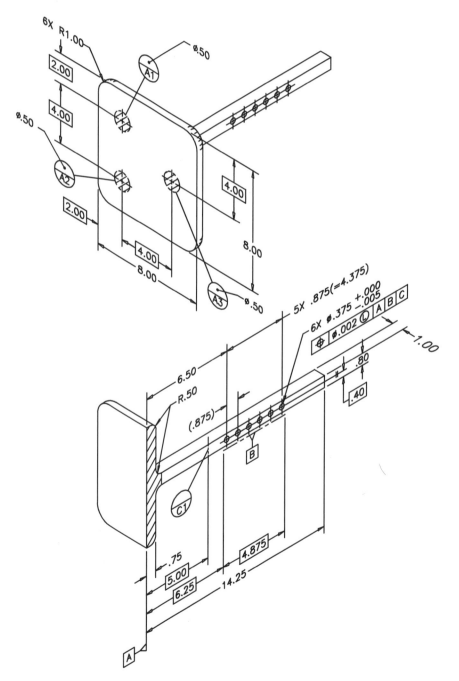

DRAFTING PROBLEM 18

Units
METRIC

Application
Entry level, positional tolerance for coaxial holes of different size.

Name
BRACKET

Material
SAE 1030

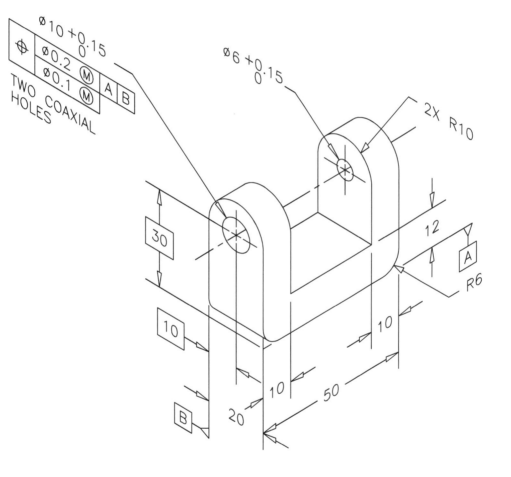

DRAFTING PROBLEM 19

Units
INCH

Application
Advanced level, positional tolerance for coaxial holes of the same size. The object in the problem is displayed in full section for clarity.

Name
BEARING GUIDE

Material
SAE 2340

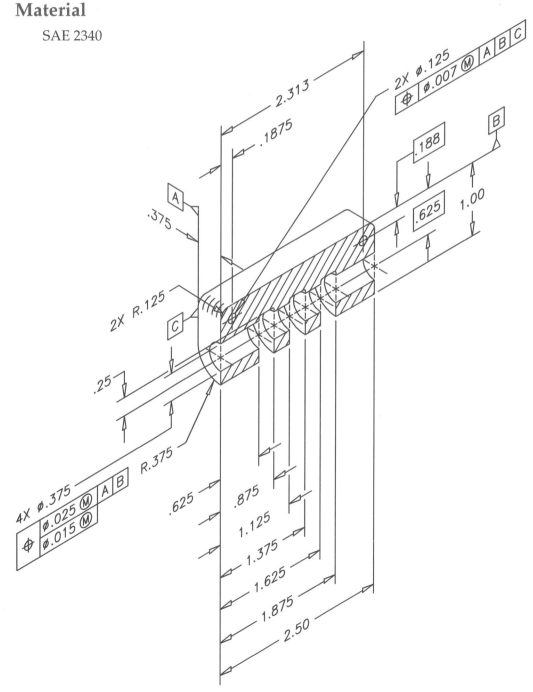

DRAFTING PROBLEM 20

Units
METRIC

Application
Medium level, positional tolerance with two single-segment feature control frame.

Name
KEYED STOP PLATE

Material
SAE 1045

Finish
ALL OVER 0.5μM

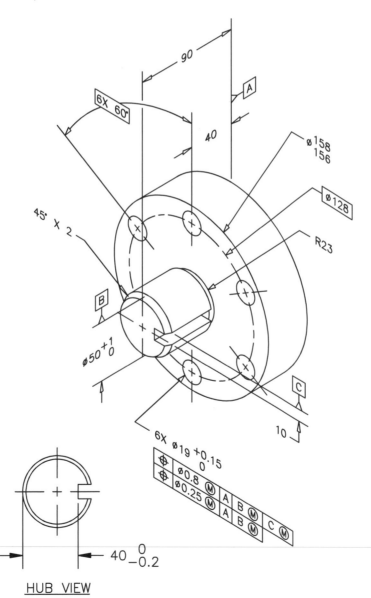

HUB VIEW

DRAFTING PROBLEM 21

Units
INCHES

Application
Advanced level, multiple positional tolerance with two single-segment feature control frames for a pattern of features.

Name
CLUTCH PLATE

Material
SAE 1060

Finish
ALL OVER 0.5μM

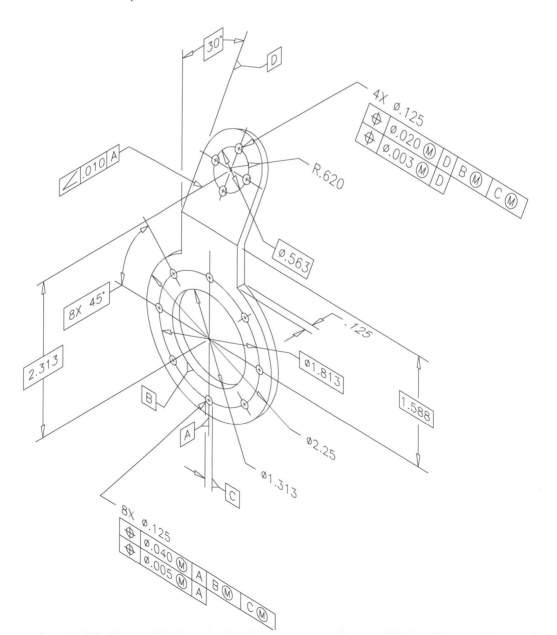

DRAFTING PROBLEM 22

Units
METRIC

Application
Medium level, positional tolerance for coaxial features, zero geometric tolerance at MMC.

Name
MOUNTING PLATE

Material
SAE 4140

Finish
ALL OVER 0.8μM

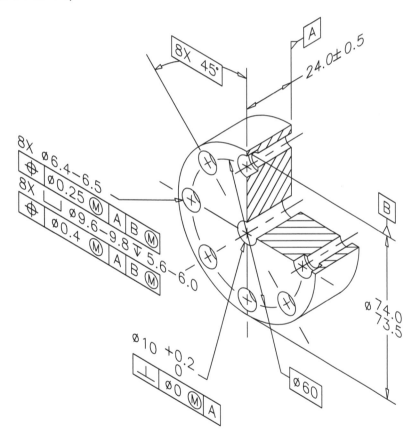

DRAFTING PROBLEM 23

Units
METRIC

Application
Medium level; positional tolerance for a pattern, for symmetry, and for perpendicularity.

Name
MOUNTING PLATE

Material
SAE 1095

Finish
ALL OVER 0.2μM

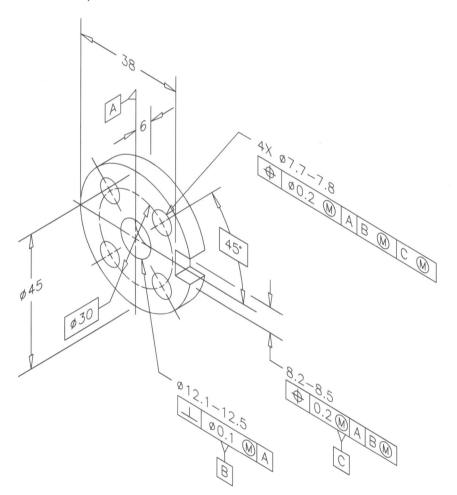

DRAFTING PROBLEM 24

Units
METRIC

Application
Medium level, positional tolerance for holes with separate requirements.

Name
SPRING CLIP

Material
SAE 1085

Finish
ALL OVER 0.2μM

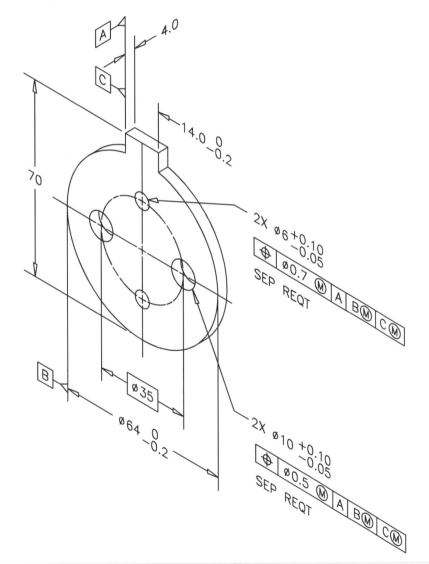

DRAFTING PROBLEM 25

Units
METRIC

Application
Medium level, least material condition applied to a pattern of slots.

Name
SPLINE COLLAR

Material
SAE 3140

Finish
ALL OVER 0.2µM

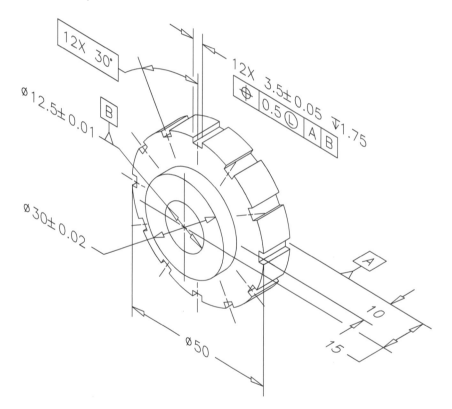

DRAFTING PROBLEM 26

Units
METRIC

Application
Medium level, positional tolerance for the symmetry of tabs.

Name
LOCKING COLLAR

Material
SAE 1080

Finish
ALL OVER 0.2µM

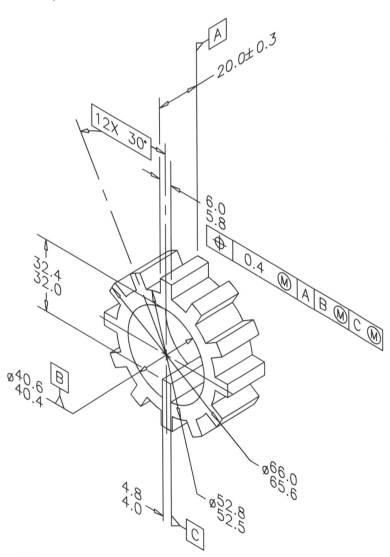

DRAFTING PROBLEM 27

Units
INCHES

Application
Entry level, projected tolerance zone.

Name
THREADED PLATE

Material
SAE 1020

Finish
ALL OVER 0.8μM

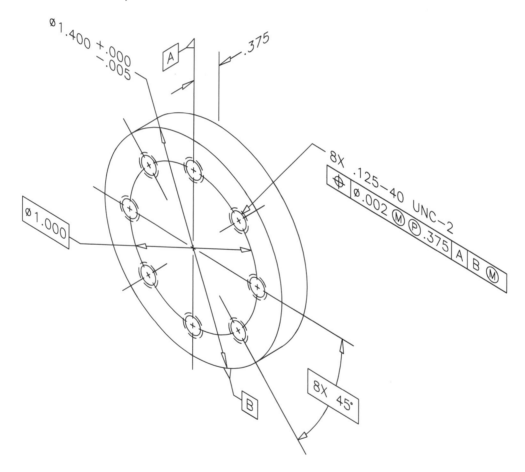

DRAFTING PROBLEM 28

Units
METRIC

Application
Advanced level, positional tolerance of nonparallel features, datum target symbols establishing datum axis, working from a rough engineering sketch.

Name
OSCILLATOR

Material
PHOSPHOR BRONZE

Finish
ALL OVER 0.25μM

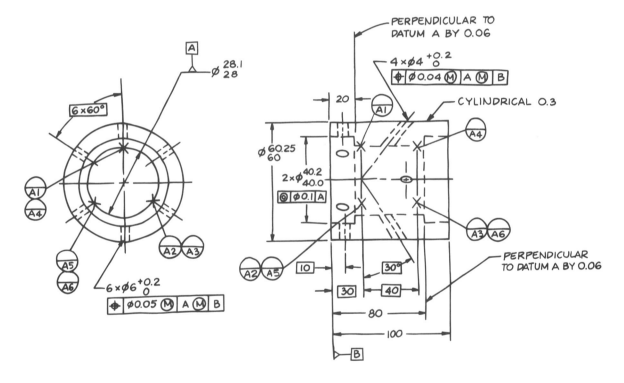

DRAFTING PROBLEM 29

Units
INCHES

Application
Advanced level, multiple applications from written instructions.

Name
END BRACKET

Material
TITANIUM

Finish
63µIN

Additional Instructions
1. Label all datums.
2. Use the dimensions shown on the sketch, but not necessarily the placement shown.
3. Provide an angularity tolerance of .005 to Datum C for the 70° angle both sides.
4. Hold surface profile of .005 between Point X and Point Y at both surfaces controlled by the R1.25 dimension.
5. Position the Ø.750+.005/−.002 and the counterbore Ø1.380 to datums A, B, and C by .003 MMC.
6. The surfaces labeled 1, 2, 3, and 4 shall be held symmetrical with datum center plane D by .004. (Note the numbers 1, 2, 3, and 4 will not be shown on your drawing.)
7. Provide a coaxial position tolerance to locate the Ø.187 feature with a coaxial Ø tolerance of .002 at MMC relative to datums A, B, and C where the holes (together) must lie. Also, control the coaxial Ø.001 tolerance at MMC where the axes of the holes must lie relative to each other.
8. Provide reference to Datum E. Datum E is surface number 1.
9. Provide a coaxial position tolerance to locate the Ø.86–.89 feature with a coaxial Ø tolerance of .005 at MMC relative to datums E, A, and C where the holes (together) must lie. Also, control the coaxial Ø.002 tolerance at MMC where the axes of the holes must lie relative to each other.
10. Establish Datum B perpendicular to Datum A by .002.
11. Establish Datum C perpendicular to Datum A and Datum B by .002.

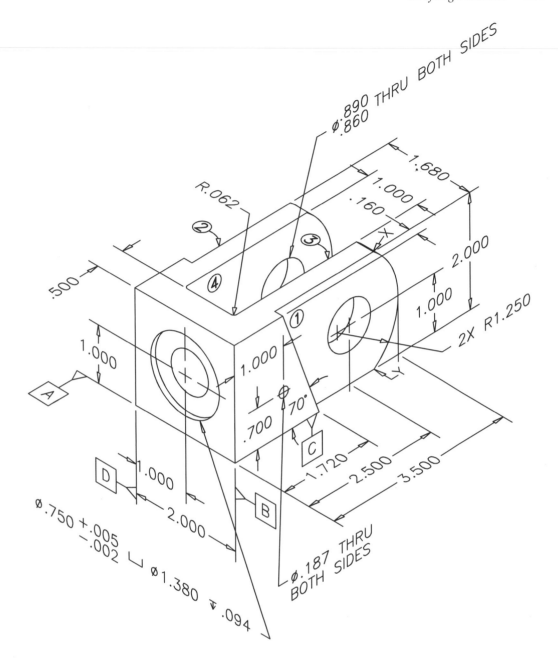

ø.890 THRU BOTH SIDES
.860

R.062

1.680

1.000

.160

2.000

1.000

.500

2X R1.250

1.000

1.000

A

.700

70°

C

1.000

D

B

1.720

2.500

3.500

2.000

ø.750 +.005 / −.002 ⊔ ø1.380 ⊽ .094

ø.187 THRU BOTH SIDES

DRAFTING PROBLEM 30

Units
METRIC

Application
Medium level, projected tolerance zone.

Name
TRANSMISSION COVER

Material
CAST IRON (CI)

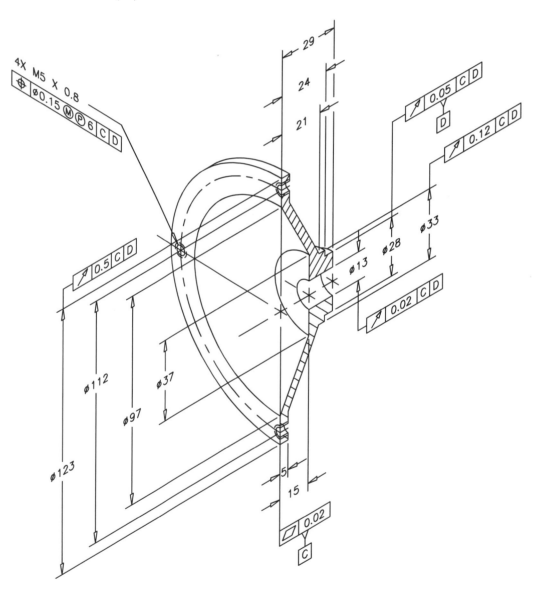

SLEEVE-DEWAR REIMAGING

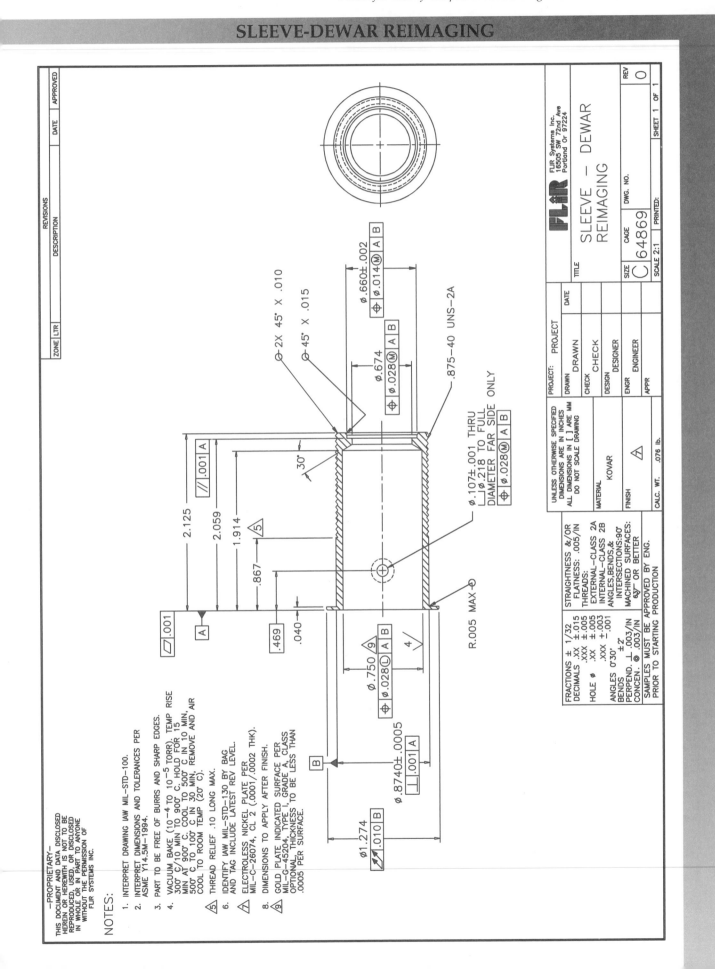

NOTES:

—PROPRIETARY—
THIS DOCUMENT AND DATA DISCLOSED
HEREIN OR HEREWITH IS NOT TO BE
REPRODUCED, USED, OR DISCLOSED
IN WHOLE OR IN PART TO ANYONE
WITHOUT THE PERMISSION OF
FLIR SYSTEMS INC.

1. INTERPRET DRAWING IAW MIL-STD-100.

2. INTERPRET DIMENSIONS AND TOLERANCES PER
   ASME Y14.5M-1994.

3. PART TO BE FREE OF BURRS AND SHARP EDGES.

4. VACUUM BAKE (10^{-4} TO 10^{-5} TORR). TEMP RISE
   300° C/10 MIN TO 900° C. HOLD FOR 15
   MIN AT 900° C. COOL TO 500° C IN 10 MIN,
   500° C TO 100° C IN 30 MIN, REMOVE AND AIR
   COOL TO ROOM TEMP (20° C).

5. THREAD RELIEF .10 LONG MAX.

6. IDENTIFY IAW MIL-STD-130 BY BAG
   AND TAG INCLUDE LATEST REV LEVEL.

7. ELECTROLESS NICKEL PLATE PER
   MIL-C-26074, CL 2 (.0001/.0002 THK).

8. DIMENSIONS TO APPLY AFTER FINISH.

9. GOLD PLATE INDICATED SURFACE PER
   MIL-G-45204, TYPE I, GRADE A, CLASS
   OPTIONAL THICKNESS TO BE LESS THAN
   .0005 PER SURFACE.

FLIR Systems Inc.
16505 SW 72nd Ave
Portland Or 97224

TITLE   SLEEVE – DEWAR
         REIMAGING

SIZE CAGE   DWG. NO.
C   64869

SCALE 2:1   PRINTED: SHEET 1 OF 1

REV 0

PROJECT:	PROJECT		DATE
DRAWN	DRAWN		
CHECK	CHECK		
DESIGN	DESIGNER		
ENGR	ENGINEER		
APPR	APPR		

UNLESS OTHERWISE SPECIFIED
DIMENSIONS ARE IN INCHES
ALL DIMENSIONS IN [ ] ARE MM
DO NOT SCALE DRAWING

MATERIAL   KOVAR

FINISH

CALC. WT.   .076 Ib.

STRAIGHTNESS &/OR
FLATNESS: .005/IN
THREADS:
EXTERNAL–CLASS 2A
INTERNAL–CLASS 2B
ANGLES,BENDS,&
INTERSECTIONS:90°
MACHINED SURFACES:
63 OR BETTER

FRACTIONS ± 1/32
DECIMALS .XX   ±.015
        .XXX ±.005
HOLE ø .XX   ±.005
        .XXX ±.001
ANGLES 0°30'
BENDS   ±2°
PERPEND. ⊥ .003/IN
CONCEN. ◎ .003/IN

SAMPLES MUST BE APPROVED BY ENG.
PRIOR TO STARTING PRODUCTION

REVISIONS
ZONE LTR   DESCRIPTION   DATE APPROVED

ø.660±.002
⊕ ø.014 Ⓜ A B
ø.674
⊕ ø.028 Ⓜ A B
.875–40 UNS–2A

2X 45° X .010
45° X .015

// .001 A
2.125
2.059
1.914
.867
30°
⬦ .001
A

ø.107±.001 THRU
⌴ø.218 TO FULL
DIAMETER FAR SIDE ONLY
⊕ ø.028 Ⓜ A B

.469
.040

ø.750 9
⊕ ø.028 Ⓛ A B
4

R.005 MAX

⌖ .001 A

ø.8740±.0005
⟂ .001 A

B

ø1.274 B
⟋ .010 B

BRACKET

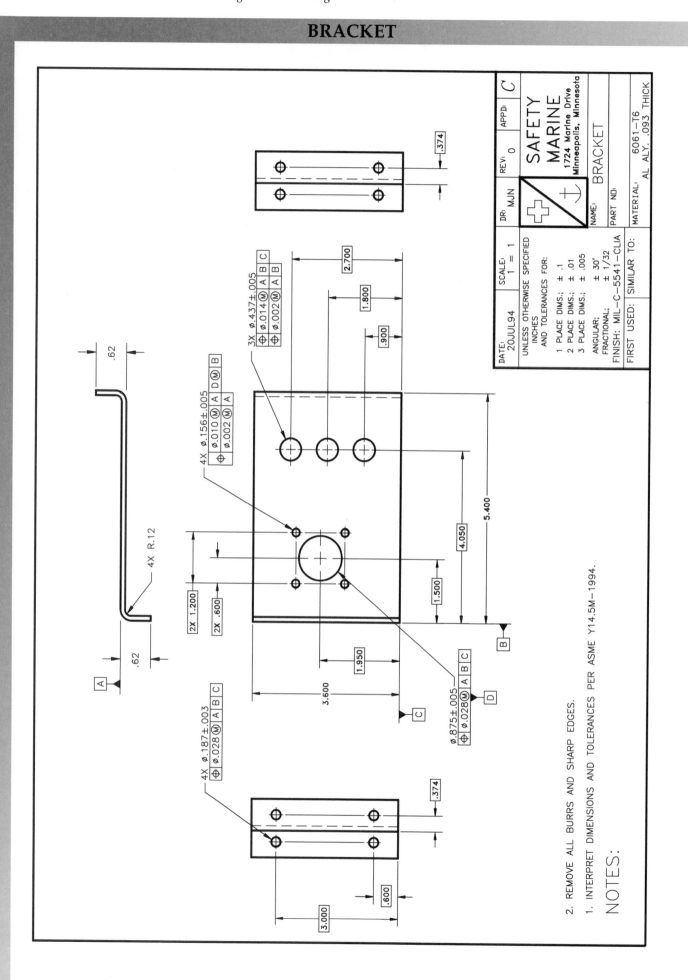

HUB-STATIONARY, ATU

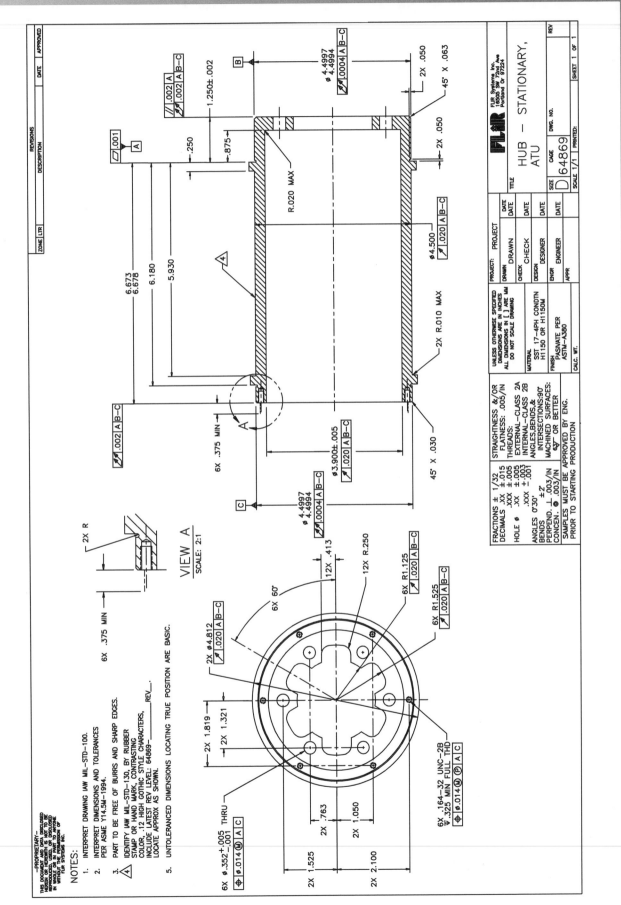

NOTES:

1. INTERPRET DRAWING IAW MIL-STD-100.

2. INTERPRET DIMENSIONS AND TOLERANCES PER ASME Y14.5M-1994.

3. PART TO BE FREE OF BURRS AND SHARP EDGES.

4. IDENTIFY IAW MIL-STD-130, BY RUBBER STAMP OR HAND MARK, CONTRASTING COLOR, .12 HIGH GOTHIC STYLE CHARACTERS, INCLUDE LATEST REV LEVEL: 64869-___ REV ___. LOCATE APPROX AS SHOWN.

5. UNTOLERANCED DIMENSIONS LOCATING TRUE POSITION ARE BASIC.

VIEW A
SCALE: 2:1

PEDAL-ACCELERATOR

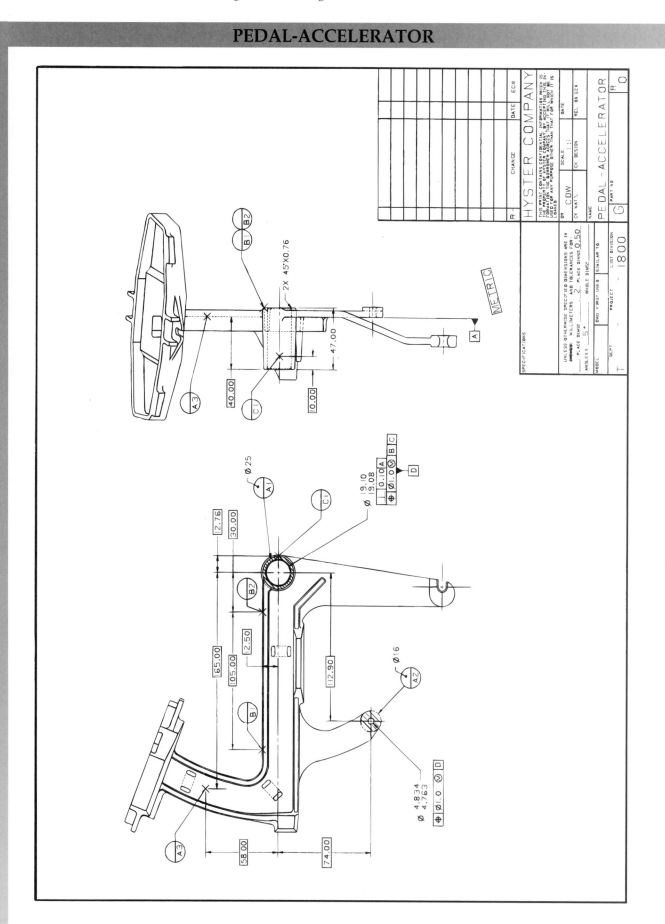

MOUNTING PLATE (UPPER)-FRAME ASSY 3 AXIS HP

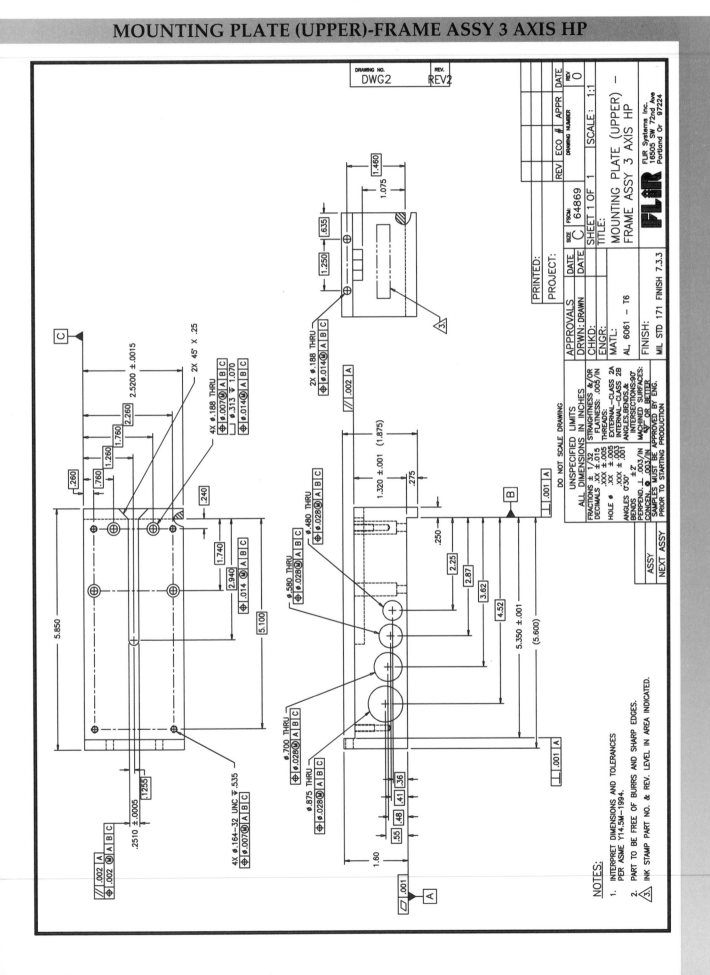

HYDRAULIC VALVE

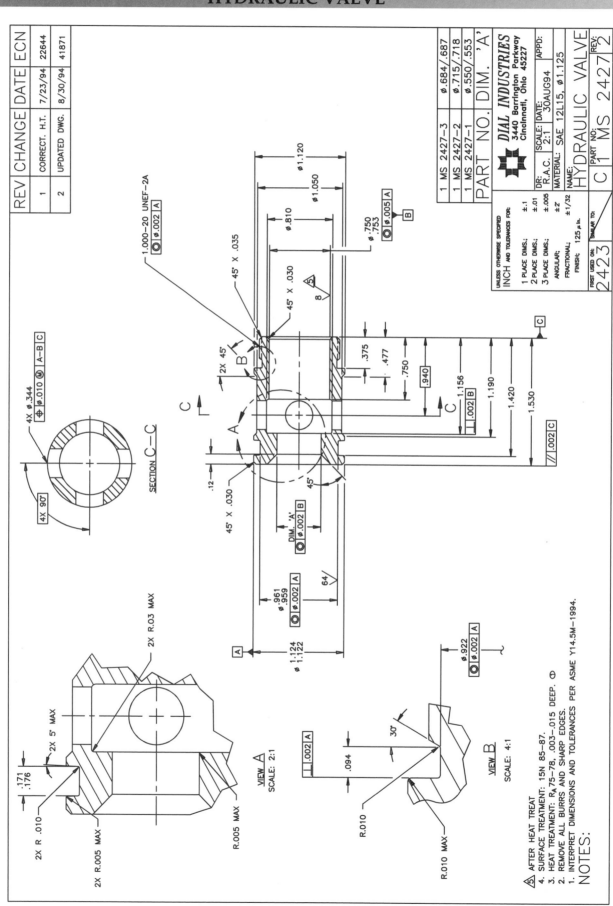

COVER, CAGE-INNER AZ DRIVE

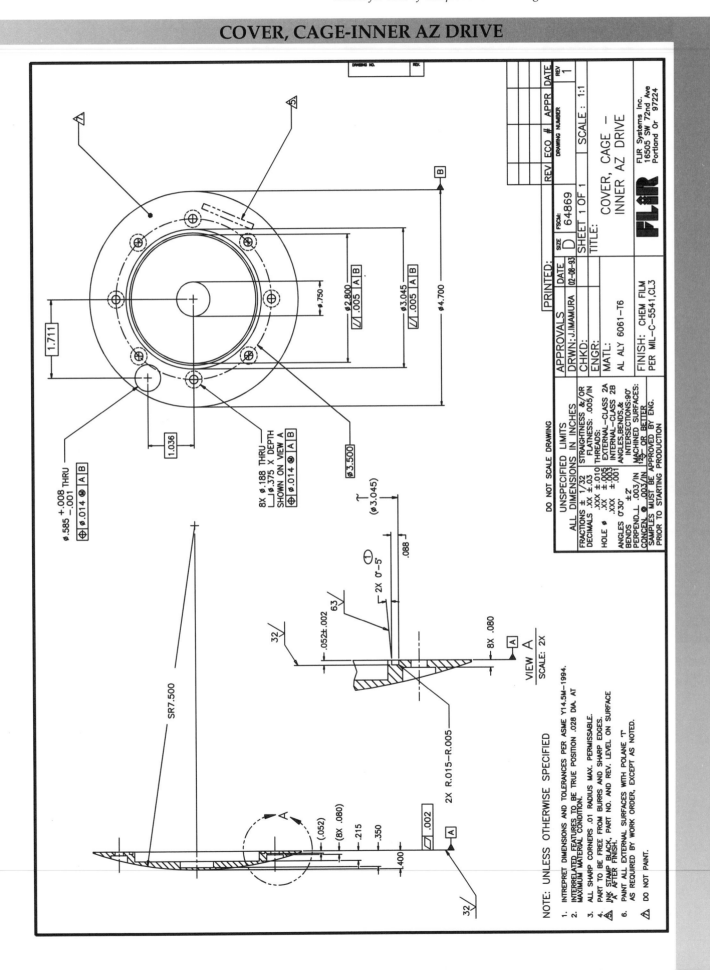

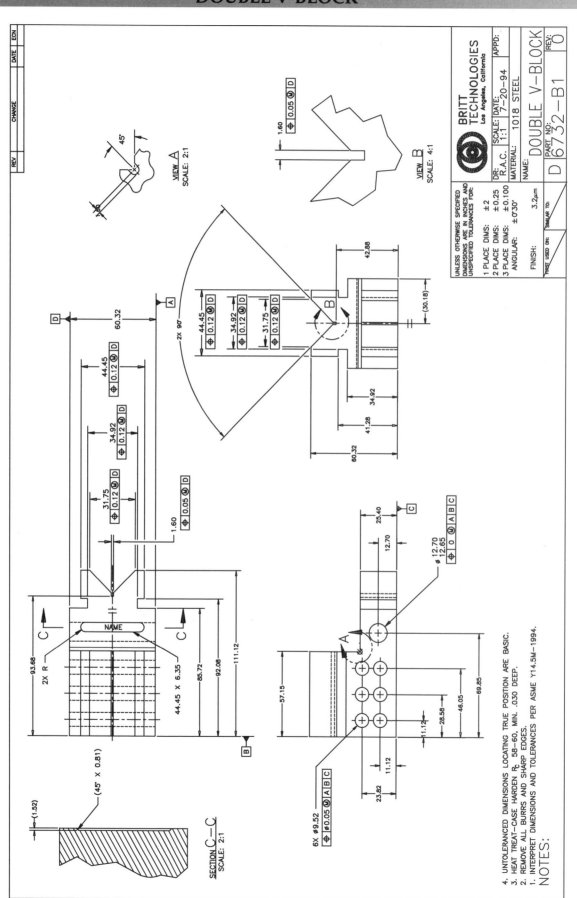

PLATE-TOP MOUNTING

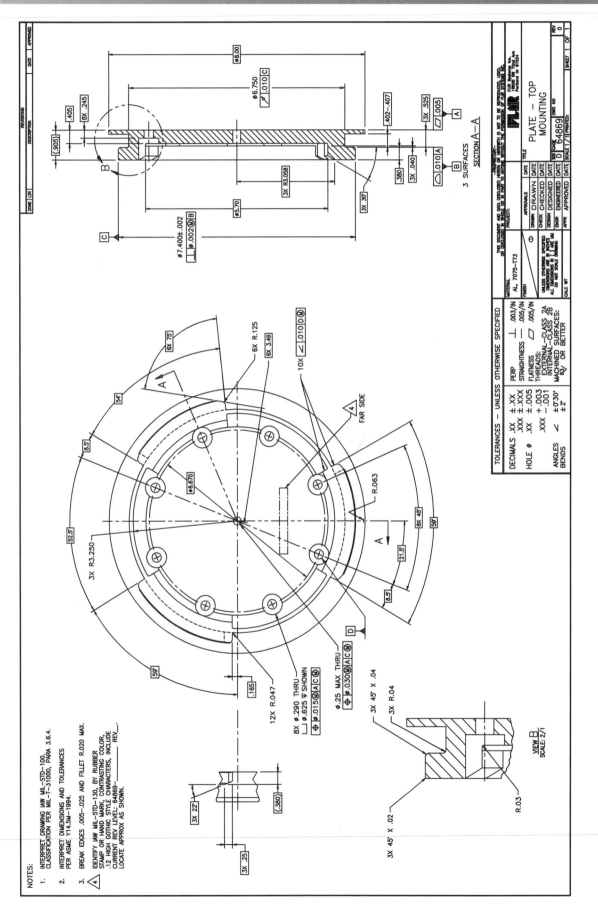

B.H.-TOP

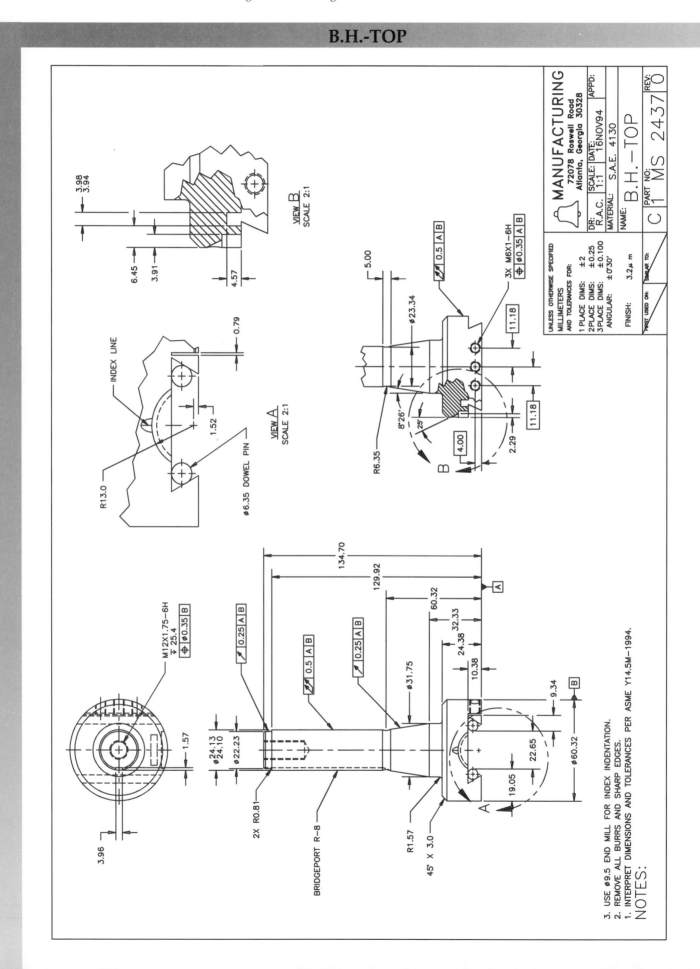

HOUSING-LENS, FOCUS

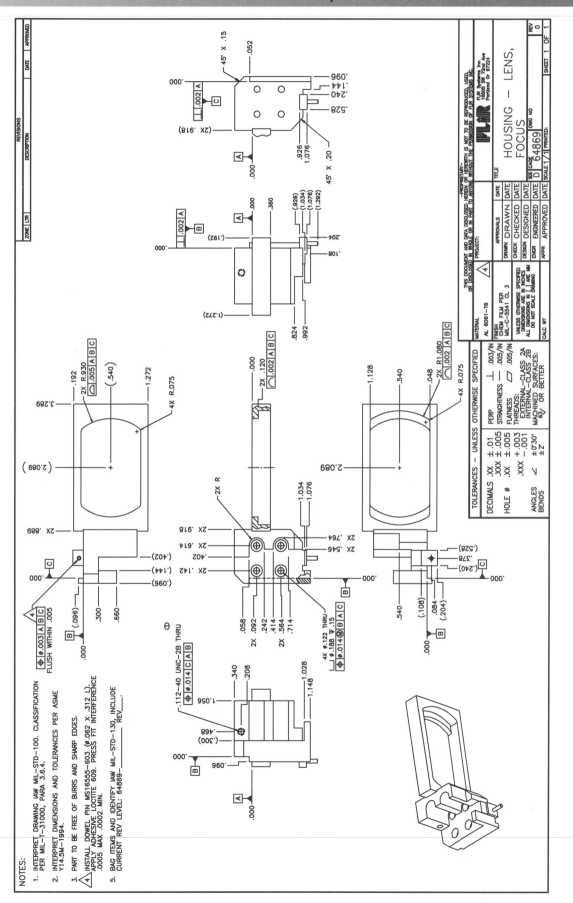

PLATE-BOTTOM WEDGED, ADJUSTABLE PARALLEL (HP)

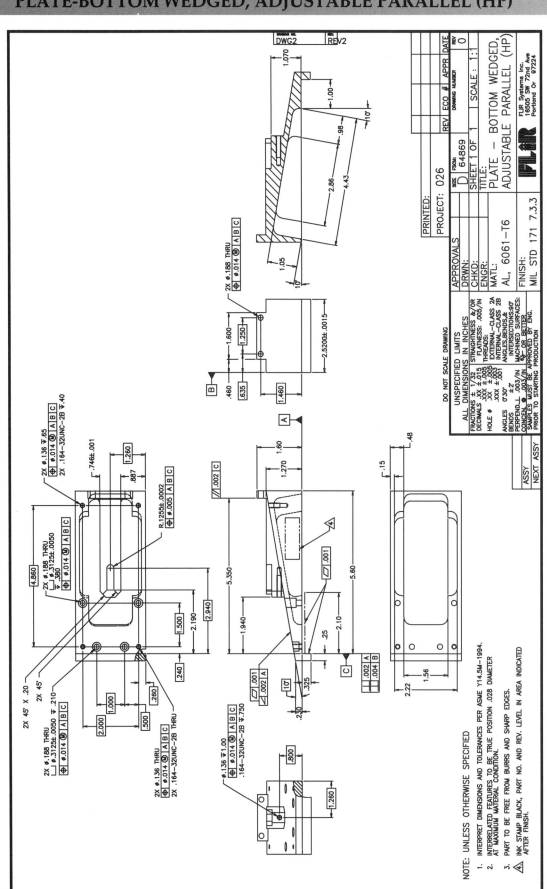

BRACKET ASSY-EL GIMBAL

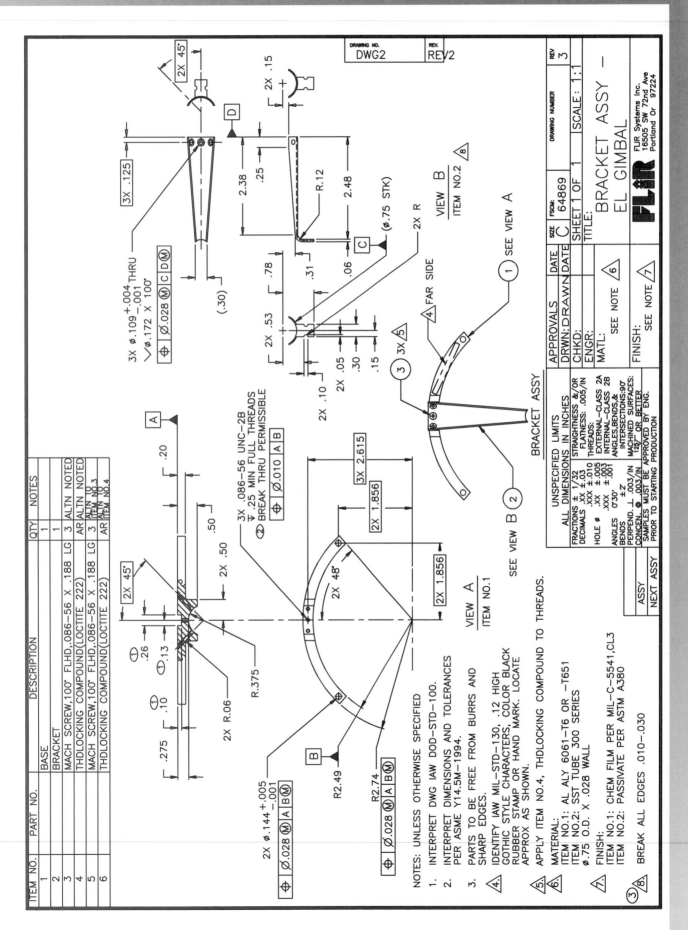

Appendix
Dimensioning Symbols A1

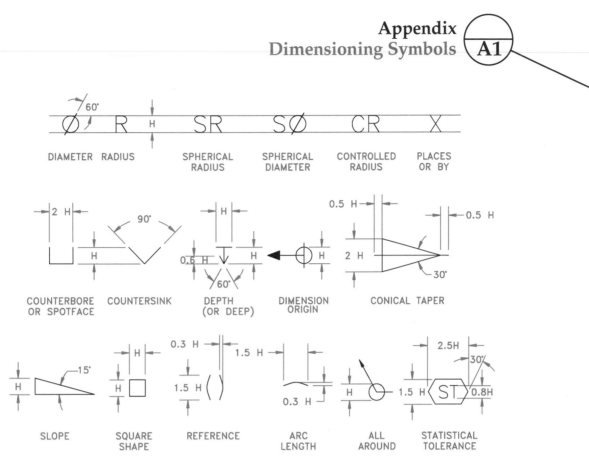

DIMENSIONING SYMBOLS

Appendix
Datum Feature and Datum Target Symbols

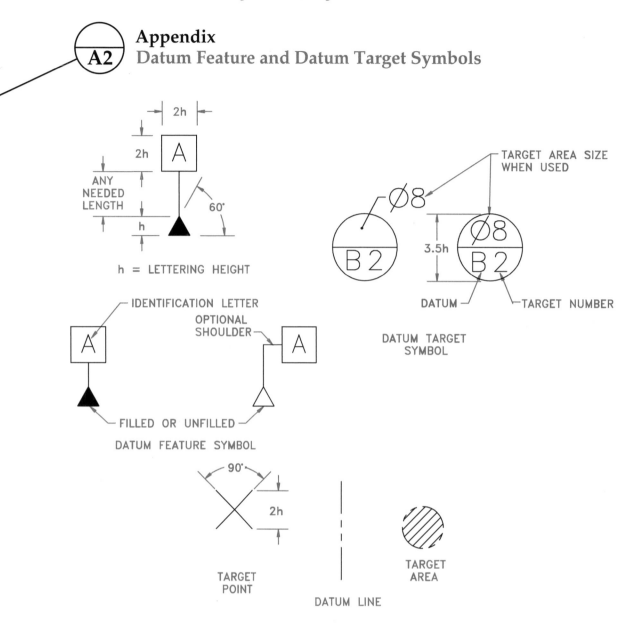

h = LETTERING HEIGHT

IDENTIFICATION LETTER

OPTIONAL SHOULDER

FILLED OR UNFILLED

DATUM FEATURE SYMBOL

TARGET AREA SIZE WHEN USED

DATUM

TARGET NUMBER

DATUM TARGET SYMBOL

TARGET POINT

DATUM LINE

TARGET AREA

DATUM FEATURE AND DATUM TARGET SYMBOLS

Appendix
Material Condition Symbols A3

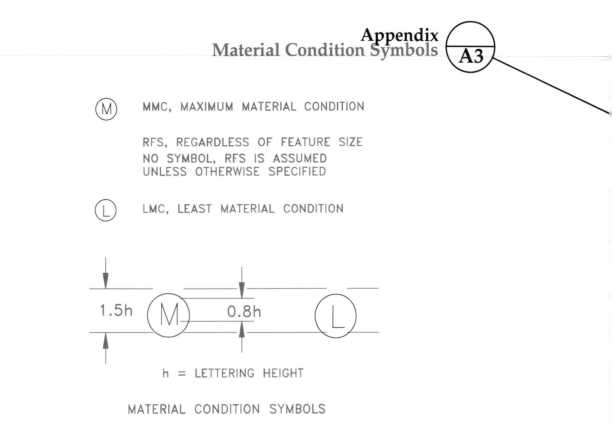

Ⓜ   MMC, MAXIMUM MATERIAL CONDITION

RFS, REGARDLESS OF FEATURE SIZE
NO SYMBOL, RFS IS ASSUMED
UNLESS OTHERWISE SPECIFIED

Ⓛ   LMC, LEAST MATERIAL CONDITION

1.5h   Ⓜ   0.8h   Ⓛ

h = LETTERING HEIGHT

MATERIAL CONDITION SYMBOLS

Appendix
A4 Feature Control Frame

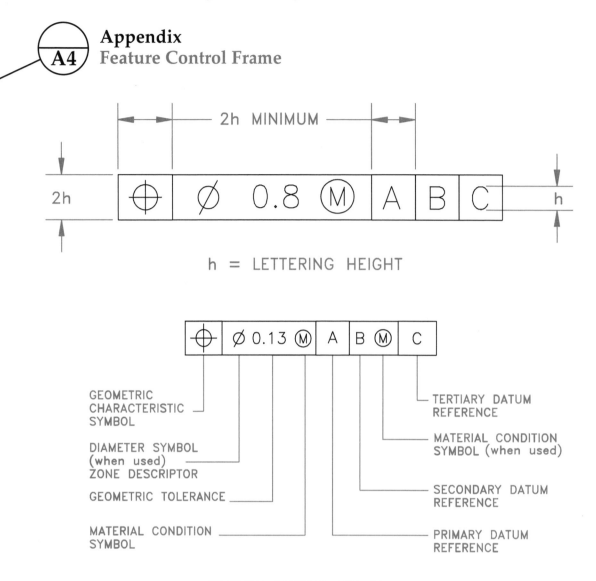

FEATURE CONTROL FRAME

Appendix
Geometric Characteristic Symbols and
Related Geometric Tolerancing Symbols

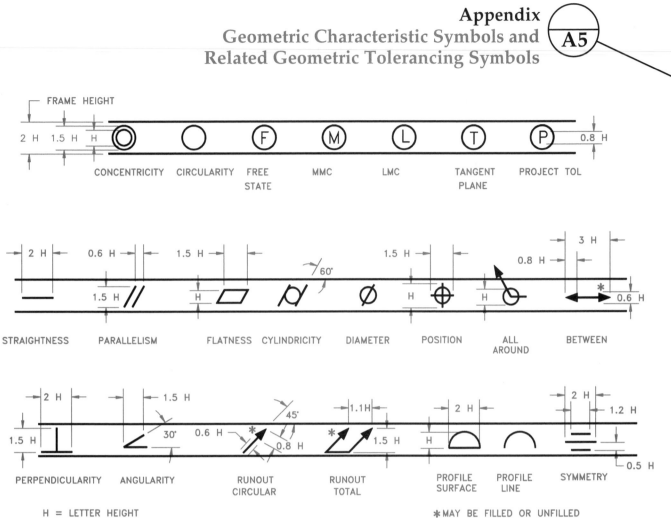

STRAIGHTNESS   PARALLELISM   FLATNESS CYLINDRICITY   DIAMETER   POSITION   ALL   BETWEEN
                                                                                                    AROUND

CONCENTRICITY CIRCULARITY FREE   MMC   LMC   TANGENT   PROJECT TOL
                           STATE   PLANE

PERPENDICULARITY   ANGULARITY   RUNOUT   RUNOUT   PROFILE   PROFILE   SYMMETRY
                                        CIRCULAR   TOTAL   SURFACE   LINE

H = LETTER HEIGHT   *MAY BE FILLED OR UNFILLED

Appendix
A6

Decimal, Fractional, Letter, Wire Gauge, and Millimeter Equivalents

EQUIVALENTS
of Regular Sizes including Decimal, Fractional, Letter, Wire Gauge, and Millimeter Sizes

Inch	Dec-imal	Wire & Letter	M/M	Inch	Dec-imal	Wire & Letter	M/M	Inch	Dec-imal	Wire & Letter	M/M	Inch	Dec-imal	Wire & Letter	M/M
	.0059	97	.15		.0250	72			.0670	51		1/8	.1220		3.1
	.0063	96	.16		.0256		.65		.0689		1.75		.1250		3.17
	.0067	95	.17		.0260	71			.0700	50			.1260		3.2
	.0071	94	.18		.0276		.7		.0709		1.8		.1280		3.25
	.0075	93	.19		.0280	70			.0728		1.85		.1285	30	
	.0079	92	.20		.0292	69			.0730	49			.1299		3.3
	.0083	91	.21		.0295		.75		.0748		1.9		.1339		3.4
	.0087	90	.22		.0310	68			.0760	48			.1360	29	
	.0089			1/32	.0312		.79	5/64	.0768		1.95		.1378		3.5
	.0091	89	.23		.0315		.8		.0781		1.98		.1405	28	
	.0094		.24		.0320	67			.0785	47		9/64	.1406		3.57
	.0095	88			.0330	66			.0787		2.		.1417		3.6
	.0100	87	.25		.0335		.85		.0807		2.05		.1440	27	
	.0102		.26		.0350	65			.0810	46			.1457		3.7
	.0105	86			.0354		.9		.0820	45			.1470	26	
	.0106		.27		.0360	64			.0827		2.1		.1476		3.75
	.0110	85	.28		.0370	63			.0846		2.15		.1495	25	
	.0114		.29		.0374		.95		.0860	44			.1496		3.8
	.0115	84			.0380	62			.0866		2.2		.1520	24	
	.0118		.30		.0390	61			.0886		2.25		.1535		3.9
	.0120	83			.0394		1.		.0890	43			.1540	23	
	.0125	82			.0400	60			.0906		2.3	5/32	.1562		3.97
	.0126		.32		.0410	59			.0925		2.35		.1570	22	
	.0130	81			.0413		1.05		.0935	42			.1575		4.
	.0134		.34		.0420	58		3/32	.0938		2.38		.1590	21	
	.0135	80			.0430	57			.0945		2.4		.1610	20	
	.0138		.35		.0433		1.1		.0960	41			.1614		4.1
	.0142		.36		.0453		1.15		.0965		2.45		.1654		4.2
	.0145	79		3/64	.0465	56			.0980	40			.1660	19	
1/64	.0150		.38		.0469		1.19		.0984		2.5		.1673		4.25
	.0156		.397		.0472		1.2		.0995	39			.1693		4.3
	.0157		.4		.0492		1.25		.1015	38			.1695	18	
	.0160	78			.0512		1.3		.1024		2.6	11/64	.1719		4.37
	.0165		.42		.0520	55			.1040	37			.1730	17	
	.0173		.44		.0531		1.35		.1063		2.7		.1732		4.4
	.0177		.45		.0550	54			.1065	36			.1770	16	
	.0180	77			.0551		1.4	7/64	.1083		2.75		.1772		4.5
	.0181		.46		.0571		1.45		.1094		2.78		.1800	15	
	.0189		.48		.0591		1.5		.1100	35			.1811		4.6
	.0197		.5		.0595	53			.1102		2.8		.1820	14	
	.0200	76			.0610		1.55		.1110	34			.1850	13	
	.0210	75		1/16	.0625		1.59		.1130	33			.1850		4.7
	.0217		.55		.0630		1.6		.1142		2.9		.1870		4.75
	.0225	74			.0635	52			.1160	32	3/16	.1875		4.76
	.0236		.6		.0650		1.65		.1181		3.		.1890		4.8
	.0240	73			.0669		1.7		.1200	31			.1890	12	

Rutland Tool and Supply Co., Inc.

Appendix A6
Decimal, Fractional, Letter, Wire Gauge, and Millimeter Equivalents

EQUIVALENTS
of Regular Sizes including Decimal, Fractional, Letter, Wire Gauge, and Millimeter Sizes

Inch	Decimal	Wire & Letter	M/M	Inch	Decimal	Wire & Letter	M/M	Inch	Decimal	Wire & Letter	M/M	Inch	Decimal	Wire & Letter	M/M
	.1910	11			.2677		6.8		.3622		9.2	41/64	.6406		16.28
	.1929		4.9		.2717		6.9		.3642		9.25		.6496		16.5
	.1935	10			.2720	I			.3661		9.3	21/32	.6562		16.67
	.1960	9			.2756		7.		.3680	U			.6693		17.
	.1969		5.		.2770	J			.3701		9.4	43/64	.6719		17.07
	.1990	8			.2795		7.1		.3740		9.5	11/16	.6875		17.46
	.2008		5.1		.2810	K		3/8	.3750				.6890		17.5
	.2010	7		9/32	.2813		7.14		.3770	V		45/64	.7031		17.86
13/64	.2031		5.15		.2835		7.2		.3780		9.6		.7087		18.
	.2040	6			.2854		7.25		.3819		9.7	23/32	.7188		18.26
	.2047		5.2		.2874		7.3		.3839		9.75		.7283		18.5
	.2055	5			.2900	L			.3858		9.8	47/64	.7344		18.65
	.2067		5.25		.2913		7.4		.3860	W			.7480		19.
	.2087		5.3		.2950	M			.3898		9.9	3/4	.7500		19.05
	.2090	4			.2953		7.5	25/64	.3906			49/64	.7656		19.45
	.2126		5.4	19/64	.2969		7.54		.3937		10.		.7677		19.5
7/32	.2130	3	5.41		.2992		7.6		.3970	X		25/32	.7812		19.84
	.2165		5.5		.3020	N			.4040	Y			.7874		20.
	.2187		5.56		.3031		7.7	13/32	.4062			51/64	.7969		20.24
	.2205		5.6		.3051		7.75		.4130	Z			.8071		20.5
	.2210	2	5.61		.3071		7.8		.4134		10.5	13/16	.8125		20.64
	.2244		5.7		.3110		7.9	27/64	.4219				.8268		21.
	.2264		5.75	5/16	.3125		7.94		.4331		11.	53/64	.8281		21.03
	.2280	1	5.79		.3150		8.	7/16	.4375			27/32	.8438		21.43
	.2283		5.8		.3160	O			.4528		11.5		.8465		21.5
	.2323		5.9		.3189		8.1	29/64	.4531				.8594		21.83
	.2340	A			.3228		8.2	15/32	.4688			55/64	.8661		22.
15/64	.2344		5.95		.3230	P			.4724		12.	7/8	.8750		22.22
	.2362		6.		.3248		8.25	31/64	.4844				.8858		22.5
	.2380	B			.3268		8.3		.4921		12.5		.8906		22.62
	.2402		6.1	21/64	.3281		8.33	1/2	.5000			57/64	.9055		23.
	.2420	C			.3307		8.4		.5118		13.	29/32	.9062		23.02
	.2441		6.2		.3320	Q		33/64	.5156				.9219		23.42
	.2460	D			.3346		8.5	17/32	.5312				.9252		23.5
	.2461		6.25		.3386		8.6		.5315		13.5	15/16	.9375		23.81
	.2480		6.3		.3390	R		35/64	.5469				.9449		24.
1/4	.2500	E	6.35		.3425		8.7		.5512		14.	61/64	.9531		24.21
	.2520		6.4	11/32	.3438		8.74	9/16	.5625				.9646		24.5
	.2559		6.5		.3445		8.75		.5709		14.5	31/32	.9688		24.61
	.2570	F			.3465		8.8	37/64	.5781				.9843		25.
	.2598		6.6		.3480	S			.5906		15.	63/64	.9844		25.01
	.2610	G			.3504		8.9	19/32	.5938			1	1.0000		25.4
	.2638		6.7		.3543		9.	39/64	.6094				1.0039		25.5
17/64	.2656		6.75		.3580	T			.6102		15.5	1-1/64	1.0156		25.8
	.2657		6.75		.3583		9.1	5/8	.6250				1.0236		26.
	.2660	H		23/64	.3594		9.13		.6299		16.	1-1/32	1.0313		26.19

Appendix
A7 Decimal Equivalents and Tap Drill Sizes

THE L. S. STARRETT COMPANY / WORLD'S GREATEST TOOLMAKERS / ATHOL, MASSACHUSETTS 01331, U.S.A.

Starrett® PRECISION TOOLS
DECIMAL EQUIVALENTS AND TAP DRILL SIZES

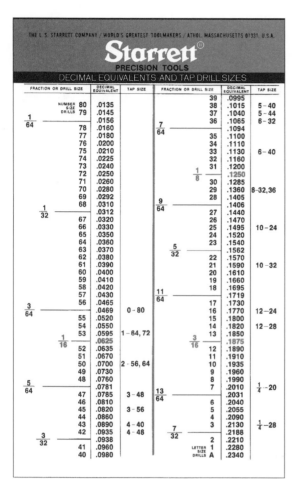

FRACTION OR DRILL SIZE	DECIMAL EQUIVALENT	TAP SIZE	FRACTION OR DRILL SIZE	DECIMAL EQUIVALENT	TAP SIZE
NUMBER SIZE DRILLS 80	.0135		39	.0995	
79	.0145		38	.1015	5-40
1/64	.0156		37	.1040	5-44
78	.0160		36	.1065	6-32
77	.0180		7/64	.1094	
76	.0200		35	.1100	
75	.0210		34	.1110	
74	.0225		33	.1130	
73	.0240		32	.1160	6-40
72	.0250		31	.1200	
71	.0260		1/8	.1250	
70	.0280		30	.1285	
69	.0292		29	.1360	8-32,36
68	.0310		28	.1405	
1/32	.0312		9/64	.1406	
67	.0320		27	.1440	
66	.0330		26	.1470	
65	.0350		25	.1495	10-24
64	.0360		24	.1520	
63	.0370		23	.1540	
62	.0380		5/32	.1562	
61	.0390		22	.1570	
60	.0400		21	.1590	10-32
59	.0410		20	.1610	
58	.0420		19	.1660	
57	.0430		18	.1695	
56	.0465		11/64	.1719	
3/64	.0469	0-80	17	.1730	
55	.0520		16	.1770	12-24
54	.0550		15	.1800	
53	.0595	1-64, 72	14	.1820	12-28
1/16	.0625		13	.1850	
52	.0635		3/16	.1875	
51	.0670		12	.1890	
50	.0700	2-56, 64	11	.1910	
49	.0730		10	.1935	
48	.0760		9	.1960	
5/64	.0781		8	.1990	
47	.0785	3-48	7	.2010	1/4-20
46	.0810		13/64	.2031	
45	.0820	3-56	6	.2040	
44	.0860		5	.2055	
43	.0890	4-40	4	.2090	
42	.0935	4-48	3	.2130	1/4-28
3/32	.0938		7/32	.2188	
41	.0960		2	.2210	
40	.0980		LETTER SIZE DRILLS 1	.2280	
			A	.2340	

FRACTION OR DRILL SIZE	DECIMAL EQUIVALENT	TAP SIZE	FRACTION OR DRILL SIZE	DECIMAL EQUIVALENT	TAP SIZE
15/64	.2344		19/32	.5938	
LETTER SIZE DRILLS B	.2380		39/64	.6094	
C	.2420		5/8	.6250	
D	.2460		41/64	.6406	
1/4	.2500		21/32	.6562	3/4-10
E	.2570	5/16-18	43/64	.6719	
F	.2570		11/16	.6875	3/4-16
G	.2610		45/64	.7031	
17/64	.2656		23/32	.7188	
H	.2660		47/64	.7344	
I	.2720	5/16-24	3/4	.7500	7/8-9
J	.2770		49/64	.7656	
K	.2810		25/32	.7812	
9/32	.2812		51/64	.7969	7/8-14
L	.2900		13/16	.8125	
M	.2950		53/64	.8281	
19/64	.2969		27/32	.8438	
N	.3020	3/8-16	55/64	.8594	
5/16	.3125		7/8	.8750	1-8
O	.3160		57/64	.8906	
P	.3230		29/32	.9062	
21/64	.3281		59/64	.9219	1-12
Q	.3320	3/8-24	15/16	.9375	
R	.3390		61/64	.9531	
11/32	.3438		31/32	.9688	
S	.3480		63/64	.9844	1 1/8-7
	.3580		1	1.0000	
23/64	.3594		1 3/64	1.0469	1 1/8-12
3/8 U	.3680	7/16-14	1 7/64	1.1094	1 1/4-7
	.3750		1 1/8	1.1250	
V	.3770		1 11/64	1.1719	1 1/4-12
W	.3860		1 17/32	1.2188	1 3/8-6
25/64	.3906	7/16-20	1 1/4	1.2500	
X	.3970		1 19/64	1.2969	1 3/8-12
Y	.4040		1 11/32	1.3438	1 1/2-6
13/32	.4062		1 3/8	1.3750	
Z	.4130		1 27/64	1.4219	1 1/2-12
27/64	.4219		1 1/2	1.5000	
7/16	.4375				
29/64	.4531	1/2-20			
15/32	.4688				
31/64	.4844	9/16-12			
1/2	.5000				
33/64	.5156	9/16-18			
17/32	.5312	5/8-11			
35/64	.5469				
9/16	.5625				
37/64	.5781	5/8-18			

PIPE THREAD SIZES

THREAD	DRILL	THREAD	DRILL
1/8-27	R	1 1/2-11 1/2	1 47/64
1/4-18	7/16	2-11 1/2	2 7/32
3/8-18	37/64	2 1/2-8	2 5/8
1/2-14	23/32	3-8	3 1/4
3/4-14	59/64	3 1/2-8	3 3/4
1-11 1/2	15/32	4-8	4 1/4
1 1/4-11 1/2	1 1/2		

Courtesy of The L. S. Starrett Company

Appendix
Tap Drill Sizes **A8**

TAP DRILL SIZES

BASED ON APPROXIMATELY 75% FULL THREAD

THREAD	DRILL	THREAD	DRILL
#0–80	3/64	1-3/4–5	1-35/64
#1–64	No. 53	1-3/4–12	1-43/64
#1–72	No. 53	2–4-1/2	1-25/32
#2–56	No. 50	2–12	1-59/64
#2–64	No. 50	2-1/4–4-1/2	2-1/32
#3–48	No. 47	2-1/2–4	2-1/4
#3–56	No. 46	2-3/4–4	2-1/2
#4–40	No. 43	3–4	2-3/4
#4–48	No. 42	**TAPER PIPE**	
#5–40	No. 38		
#5–44	No. 37	1/16–27	D
#6–32	No. 36	1/8–27	Q
#6–40	No. 33	1/4–18	7/16
#8–32	No. 29	3/8–18	9/16
#8–36	No. 29	1/2–14	45/64
#10–24	No. 25	3/4–14	29/32
#10–32	No. 21	1–11-1/2	1-9/64
#12–24	No. 16	1-1/4–11-1/2	1-31/64
#12–28	No. 14	1-1/2–11-1/2	1-47/64
1/4–20	No. 7	2–11-1/2	2-13/64
1/4–28	No. 3	2-1/2–8	2-5/8
5/16–18	F	3–8	3-1/4
5/16–24	I	3-1/2–8	3-3/4
3/8–16	5/16	4–8	4-1/4
3/8–24	Q	5–8	5-9/32
7/16–14	U	6–8	6-11/32
7/16–20	25/64	**STRAIGHT PIPE**	
1/2–12	27/64		
1/2–13	27/64	1/16–27	1/4
1/2–20	29/64	1/8–27	S
9/16–12	31/64	1/4–18	29/64
9/16–18	33/64	3/8–18	19/32
5/8–11	17/32	1/2–14	47/64
5/8–18	37/64	3/4–14	15/16
3/4–10	21/32	1–11-1/2	1-11/64
3/4–16	11/16	1-1/4–11-1/2	1-33/64
7/8–9	49/64	1-1/2–11-1/2	1-3/4
7/8–14	13/16	2–11-1/2	2-7/32
1–8	7/8	2-1/2–8	2-21/32
1–12	59/64	3–8	3-9/32
1–14	15/16	3-1/2–8	3-25/32
1-1/8–7	63/64	4–8	4-9/32
1-1/8–12	1-3/64	5–8	5-11/32
1-1/4–7	1-7/64	6–8	6-13/32
1-1/4–12	1-11/64		
1-1/2–6	1-11/32		
1-1/2–12	1-27/64		

Rutland Tool and Supply Co., Inc.

Appendix
A9 METRIC TAP DRILL SIZES

METRIC TAP DRILL SIZES

METRIC TAP SIZE	RECOMMENDED METRIC DRILL				CLOSEST RECOMMENDED INCH DRILL			
	DRILL SIZE mm	INCH EQUIVALENT	PROBABLE HOLE SIZE (Inches)	PROBABLE PERCENT OF THREAD	DRILL SIZE	INCH EQUIVALENT	PROBABLE HOLE SIZE (Inches)	PROBABLE PERCENT OF THREAD
M1.6X0.35	1.25	0.0492	0.0507	69	–	–	–	–
M1.8X0.35	1.45	0.0571	0.0586	69	–	–	–	–
M2X0.4	1.60	0.0630	0.0647	69	#52	0.0635	0.0652	66
M2.2X0.45	1.75	0.0689	0.0706	70	–	–	–	–
M2.5X0.45	2.05	0.0807	0.0826	69	#46	0.0810	0.0829	67
M3X0.5	2.50	0.0984	0.1007	68	#40	0.0980	0.1003	70
M3.5X0.6	2.90	0.1142	0.1168	68	#33	0.1130	0.1156	72
M4X0.7	3.30	0.1299	0.1328	69	#30	0.1285	0.1314	73
M4.5X0.75	3.70	0.1457	0.1489	74	#26	0.1470	0.1502	70
M5X0.8	4.20	0.1654	0.1686	69	#19	0.1660	0.1692	68
M6X1	5.00	0.1968	0.2006	70	#9	0.1960	0.1998	71
M7X1	6.00	0.2362	0.2400	70	15/64	0.2344	0.2382	73
M8X1.25	6.70	0.2638	0.2679	74	17/64	0.2656	0.2697	71
M8X1	7.00	0.2756	0.2797	69	J	0.2770	0.2811	66
M10X1.5	8.50	0.3346	0.3390	71	Q	0.3320	0.3364	75
M10X1.25	8.70	0.3425	0.3471	73	11/32	0.3438	0.3483	71
M12X1.75	10.20	0.4016	0.4063	74	Y	0.4040	0.4087	71
M12X1.25	10.80	0.4252	0.4299	67	27/64	0.4219	0.4266	72
M14X2	12.00	0.4724	0.4772	72	15/32	0.4688	0.4736	76
M14X1.5	12.50	0.4921	0.4969	71	–	–	–	–
M16X2	14.00	0.5512	0.5561	72	35/64	0.5469	0.5518	76
M16X1.5	14.50	0.5709	0.5758	71	–	–	–	–
M18X2.5	15.50	0.6102	0.6152	73	39/64	0.6094	0.6144	74
M18X1.5	16.50	0.6496	0.6546	70	–	–	–	–
M20X2.5	17.50	0.6890	0.6942	73	11/16	0.6875	0.6925	74
M20X1.5	18.50	0.7283	0.7335	70	–	–	–	–
M22X2.5	19.50	0.7677	0.7729	73	49/64	0.7656	0.7708	75
M22X1.5	20.50	0.8071	0.8123	70	–	–	–	–
M24X3	21.00	0.8268	0.8327	73	53/64	0.8281	0.8340	72
M24X2	22.00	0.8661	0.8720	71	–	–	–	–
M27X3	24.00	0.9449	0.9511	73	15/16	0.9375	0.9435	78
M27X2	25.00	0.9843	0.9913	70	63/64	0.9844	0.9914	70

Rutland Tool and Supply Co., Inc.

Appendix
Area, Temperature, Weight, and Volume Equivalents   **A10**

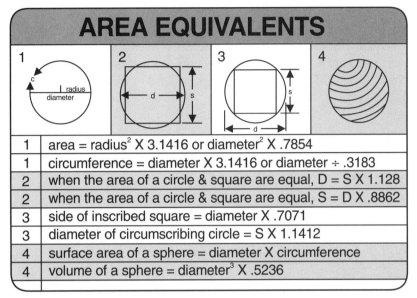

1	area = radius^2 X 3.1416 or diameter^2 X .7854
1	circumference = diameter X 3.1416 or diameter ÷ .3183
2	when the area of a circle & square are equal, D = S X 1.128
2	when the area of a circle & square are equal, S = D X .8862
3	side of inscribed square = diameter X .7071
3	diameter of circumscribing circle = S X 1.1412
4	surface area of a sphere = diameter X circumference
4	volume of a sphere = diameter^3 X .5236

Rutland Tool and Supply Co., Inc.

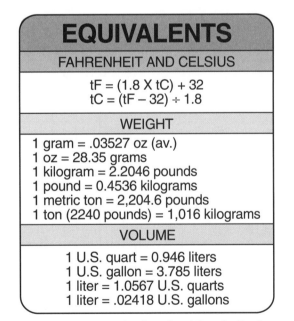

EQUIVALENTS

FAHRENHEIT AND CELSIUS

$$tF = (1.8 \times tC) + 32$$
$$tC = (tF - 32) \div 1.8$$

WEIGHT

1 gram = .03527 oz (av.)
1 oz = 28.35 grams
1 kilogram = 2.2046 pounds
1 pound = 0.4536 kilograms
1 metric ton = 2,204.6 pounds
1 ton (2240 pounds) = 1,016 kilograms

VOLUME

1 U.S. quart = 0.946 liters
1 U.S. gallon = 3.785 liters
1 liter = 1.0567 U.S. quarts
1 liter = .02418 U.S. gallons

Rutland Tool and Supply Co., Inc.

Appendix
A11 Metric Coordinate to Positional Tolerance Conversion

This chart allows you to make an approximate conversion from conventional coordinate tolerance location dimensions to a positional tolerance. To make a conversion, find the conventional coordinate tolerance along the horizontal scale at the bottom of the chart such, as the 0.1 shown with an arrow. Follow the line up to the colored diagonal line and then follow the colored arc to the positional tolerance value found along the vertical scale at the left. The positional tolerance for the 0.1 coordinate tolerance is 0.14, as displayed with the arrow at the left.

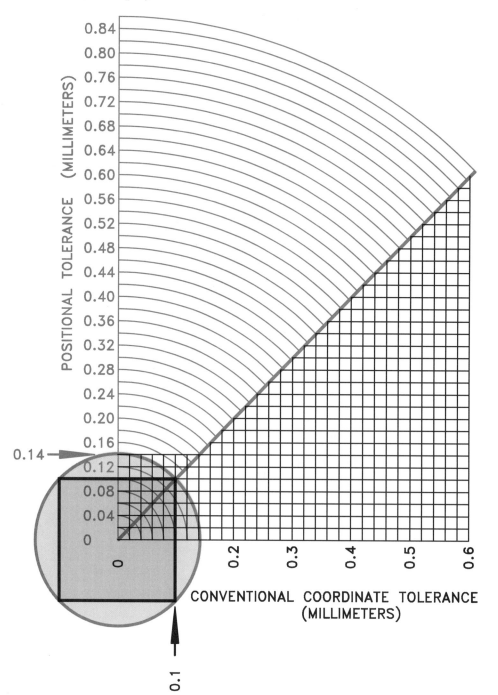

Appendix
Inch Coordinate to Positional Tolerance Conversion

A12

This chart allows you to make an approximate conversion from conventional coordinate tolerance location dimensions to a positional tolerance. To make a conversion, find the conventional coordinate tolerance along the horizontal scale at the bottom of the chart, such as the .005 shown with an arrow. Follow the line up to the colored diagonal line and then follow the colored arc to the positional tolerance value found along the vertical scale at the left. The positional tolerance for the 0.1 coordinate tolerance is .007, as displayed with the arrow at the left.

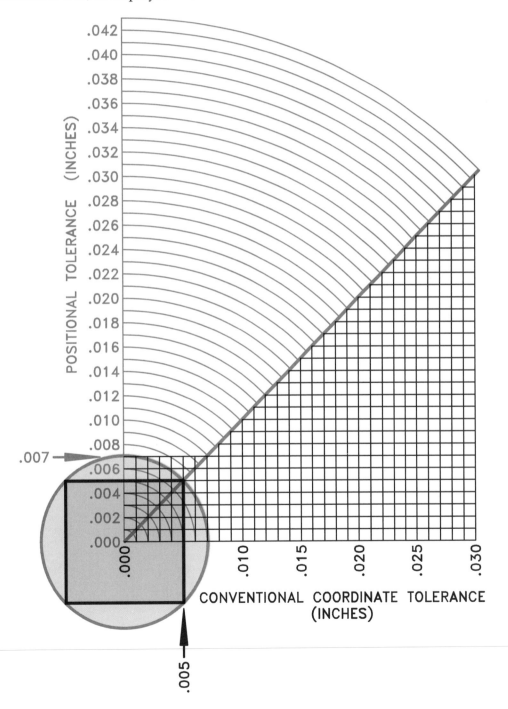

Appendix
A13 Length and Square Area Conversions

LENGTH CONVERSIONS

multiply	by	to obtain
INCHES	25.4	MILLIMETERS
FEET	304.8	MILLIMETERS
INCHES	2.54	CENTIMETERS
FEET	30.48	CENTIMETERS
MILLIMETERS	.03937008	INCHES
CENTIMETERS	.3937008	INCHES
METERS	39.37008	INCHES
MILLIMETERS	.003280840	FEET
CENTIMETERS	.03280840	FEET
INCHES	.0254	METERS

Rutland Tool and Supply Co., Inc.

SQUARE AREA CONVERSIONS

multiply	by	to obtain
MILLIMETERS	.00001076391	FEET
MILLIMETERS	.00155003	INCHES
CENTIMETERS	.1550003	INCHES
CENTIMETERS	.001076391	FEET
INCHES	645.16	MILLIMETERS
INCHES	6.4516	CENTIMETERS
INCHES	.00064516	METERS
FEET	.09290304	METERS
FEET	929.0304	CENTIMETERS
FEET	92,903.04	MILLIMETERS

Rutland Tool and Supply Co., Inc.

Appendix
Standard Gage Sheet Steel A14

Manufacturers Standard Gage for Sheet Steel

Std. Gage #	Inch Thick	Lbs. Per Sq. Ft.	Std. Gage #	Inch Thick	Lbs. Per Sq. Ft.	Std. Gage #	Inch Thick	Lbs. Per Sq. Ft.	Std. Gage #	Inch Thick	Lbs. Per Sq. Ft.
3	.2391	10.00	12	.1046	4.3570	21	.0329	1.375	30	.0120	.50000
4	.2242	9.375	13	.0897	3.7500	22	.0299	1.250	31	.0105	.43750
5	.2092	8.750	14	.0747	3.1250	23	.0269	1.125	32	.0097	.40625
6	.1943	7.500	15	.0673	2.8125	24	.0239	1.000	33	.0090	.37500
7	.1793	6.875	16	.0598	2.5000	25	.0209	.8750	34	.0082	.34375
8	.1644	6.250	17	.0538	2.2500	26	.0179	.7500	35	.0075	.31250
9	.1495	5.625	18	.0478	2.0000	27	.0164	.6875	36	.0067	.28125
10	.1345	5.000	19	.0418	1.7500	28	.0149	.6250	37	.0064	.26562
11	.1196	4.375	20	.0359	1.5000	29	.0135	.5625	38	.0060	.250

Rutland Tool and Supply Co., Inc.

Appendix
A15 Natural Trigonometric Functions

Degree	Sine	Cosine	Tangent	Cotangent
0	.00000	1.0000	.00000	∞
1	.01745	.9998	.01745	57.2900
2	.03490	.9994	.03492	28.6360
3	.05234	.9986	.05241	19.0810
4	.06976	.9976	.06993	14.3010
5	.08716	.9962	.08749	11.4300
6	.10453	.9945	.10510	9.5144
7	.12187	.9925	.12278	8.1443
8	.13920	.9903	.14050	7.1154
9	.15640	.9877	.15840	6.3138
10	.17360	.9848	.17630	5.6713
11	.19080	.9816	.19440	5.1446
12	.20790	.9781	.21260	4.7046
13	.22500	.9744	.23090	4.3315
14	.24190	.9703	.24930	4.0108
15	.25880	.9659	.26790	3.7321
16	.27560	.9613	.28670	3.4874
17	.29240	.9563	.30570	3.2709
18	.30900	.9511	.32490	3.0777
19	.32560	.9455	.34430	2.9042
20	.34200	.9397	.36400	2.7475
21	.35840	.9336	.38390	2.6051
22	.37460	.9272	.40400	2.4751
23	.39070	.9205	.42450	2.3559
24	.40670	.9135	.44520	2.2460
25	.42260	.9063	.46630	2.1445
26	.43840	.8988	.48770	2.0503
27	.45400	.8910	.50950	1.9626
28	.46950	.8829	.53170	1.8807
29	.48480	.8746	.55430	1.8040
30	.50000	.8660	.57740	1.7321
31	.51500	.8572	.60090	1.6643
32	.52990	.8480	.62490	1.6003
33	.54460	.8387	.64940	1.5399
34	.55920	.8290	.67450	1.4826
35	.57360	.8192	.70020	1.4281
36	.58780	.8090	.72650	1.3764
37	.60180	.7986	.75360	1.3270
38	.61570	.7880	.78130	1.2799
39	.62930	.7771	.80980	1.2349
40	.64280	.7660	.83910	1.1918
41	.65610	.7547	.86930	1.1504
42	.66910	.7431	.90040	1.1106
43	.68200	.7314	.93250	1.0724
44	.69470	.7193	.96570	1.0355
45	.70710	.7071	1.00000	1.0000
46	.71930	.6947	1.03550	.9657
47	.73140	.6820	1.07240	.9325
48	.74310	.6691	1.11060	.9004
49	.75470	.6561	1.15040	.8693

Appendix
Natural Trigonometric Functions **A15**

Degree	Sine	Cosine	Tangent	Cotangent
50	.7660	.64280	1.1918	.8391
51	.7771	.62930	1.2349	.8098
52	.7880	.61570	1.2799	.7813
53	.7986	.60180	1.3270	.7536
54	.8090	.58780	1.3764	.7265
55	.8192	.57360	1.4281	.7002
56	.8290	.55920	1.4826	.6745
57	.8387	.54460	1.5399	.6494
58	.8480	.52990	1.6003	.6249
59	.8572	.51500	1.6643	.6009
60	.8660	.50000	1.7321	.5774
61	.8746	.48480	1.8040	.5543
62	.8829	.46950	1.8807	.5317
63	.8910	.45400	1.9626	.5095
64	.8988	.43840	2.0503	.4877
65	.9063	.42260	2.1445	.4663
66	.9135	.40670	2.2460	.4452
67	.9205	.39070	2.3559	.4245
68	.9272	.37460	2.4751	.4040
69	.9336	.35840	2.6051	.3839
70	.9397	.34200	2.7475	.3640
71	.9455	.32560	2.9042	.3443
72	.9511	.30900	3.0777	.3249
73	.9563	.29240	3.2709	.3057
74	.9613	.27560	3.4874	.2868
75	.9659	.25880	3.7321	.2680
76	.9703	.24190	4.0108	.2493
77	.9744	.22500	4.3315	.2309
78	.9781	.20790	4.7046	.2126
79	.9816	.19080	5.1446	.1944
80	.9848	.17360	5.6713	.1763
81	.9877	.15640	6.3138	.1584
82	.9903	.13920	7.1154	.1405
83	.9925	.12187	8.1443	.1228
84	.9945	.10453	9.5144	.1051
85	.9962	.08716	11.4301	.0875
86	.9976	.06976	14.3007	.0699
87	.9986	.05234	19.0811	.0524
88	.9994	.03490	28.6363	.0349
89	.9998	.01745	57.2900	.0175
90	1.0000	.00000	∞	.0000

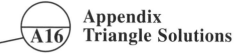

Appendix
Triangle Solutions

SOLUTIONS TO TRIANGLES

$A + B + C = 180°$

$S = \dfrac{a + b + c}{2}$

RIGHT

OBLIQUE

HAVE	WANT	FORMULAS FOR RIGHT	FORMULAS FOR OBLIQUE
abc	A	$\tan A = a/b$	$1/2A = \sqrt{(s-b)(s-c)/bc}$
	B	$90° - A$ or $\cos B = a/c$	$\sin 1/2B = \sqrt{(s-a)(s-c)/a \times c}$
	C	$90°$	$\sin 1/2C = \sqrt{(s-a)(s-b)/a \times b}$
	AREA	$a \times b/2$	$\sqrt{s \times (s-a)(s-b)(s-c)}$
aAC	B	$90° - A$	$180° - (A + C)$
	b	$a \cot A$	$a \sin B/\sin A$
	c	$a/\sin A$	$a \sin C/\sin A$
	AREA	$(a^2 \cot A)/2$	$a^2 \sin B \sin C/2 \sin A$
acC	A	$\sin A = a - c$	$\sin A = a \sin C/c$
	B	$90° - A$ or $\cos B = a/c$	$180° - (A + C)$
	b	$\sqrt{c^2 - a^2}$	$c \sin B/\sin C$
	AREA	$1/2a\sqrt{c^2 - a^2}$	$1/2\ ac \sin B$
abC	A	$\tan A = a/b$	$\tan A = a \sin C/b - a \cos c$
	B	$90° - A$ or $\tan B = b/a$	$180° - (A + C)$
	c	$\sqrt{a^2 + b^2}$	$\sqrt{a^2 + b^2 - 2ab \cos C}$
	AREA	$a \times b/2$	$1/2ab \sin C$

Rutland Tool and Supply Co., Inc.

Glossary

A

Actual Produced Size: The measured size of a feature or part after manufacturing.

Allowance: The intentional difference between the maximum material limits of mating parts.

Angularity Geometric Tolerance: A zone that is established by two parallel planes or cylindrical zones at any specified basic angle, other than 90°, to a datum plane, a pair of datum planes, or an axis.

Arrowless Dimensioning: A type of dimensioning providing only extension lines and numbers. All dimension lines and arrowheads are omitted.

Arrowless Tabular Dimensioning: Arrowless tabular dimensioning is the same as arrowless dimensioning, except identification letters are placed by the holes and the related information is given in a table.

ASME Y14.5M "Rule 1": The amount of variation in size and geometric form of a feature control. The boundary between maximum material condition (MMC) and least material condition (LMC).

ASME Y14.5M "Rule 2–All Applicable Geometric Tolerances": States that RFS applies, with respect to the individual geometric tolerance and/or datum reference, when no material condition symbol is specified.

Axis Geometric Control: The control (using geometric tolerance) of an axis of a feature or object. The feature control frame is placed with the diameter dimension of the related object or feature.

Axis Straightness: The placement of the feature control frame below the diameter dimension, and the placement of a diameter symbol in front of the geometric tolerance to specify a cylindrical tolerance zone on a drawing. This defines the control of the straightness of an axis.

B

Basic Dimension: A dimension that is considered theoretically perfect. Basic dimensions are used to describe the theoretically exact size, profile, orientation, or location of a feature or datum target.

Between: The symbol used with profile geometric tolerances to identify where the profile tolerance is applied.

Bilateral Tolerance: A tolerance permitted to vary in both the + and the – directions from the specified dimension.

C

Chain Dimensioning: A system of dimensioning where each dimension is dependent on the previous dimension.

Chain Line: A chain line is the identification of the location of the profile tolerance around the object or other characteristic related to a specific place on an object.

Chart Dimensioning: A chart that displays dimensions of the changing values of a feature. This chart provides flexibility in situations where dimensions change depending on the requirements of the product.

Circular Runout: A circular runout provides control of single circular elements of a surface. When applied to surfaces around a datum axis, circular runout controls circularity and coaxiality.

Circularity: A form tolerance that is characterized by any given cross section taken perpendicular to the axis of a cylinder or cone, or through the common center of a sphere.

Circularity Feature Control Frame: A control frame where the frame is connected with a leader to the view where the feature appears as a circle, or in the longitudinal view.

Circularity Geometric Tolerance: A tolerance that is formed by a radius zone creating two concentric circles that the actual surface must lie within.

Clearance: The loosest fit or maximum intended difference between mating parts.

Clearance Fits: These fits are generally the same as the running and sliding fits. With clearance fits, a clearance exists between the mating parts under all tolerance conditions.

Coaxial Datum Features: The establishment of a single datum axis by two datum features that are coaxial.

Coaxial Features: Features that have a common axis such as counterbores, countersinks, and counterdrills.

Coaxial Positional Tolerance: A tolerance that may be used to control the alignment of two or more holes that share a common axis.

Composite Profile Tolerance: The location of a profiled feature and, at the same time, the control of form and orientation.

Concentricity: The condition where the axes of all cross-sectional elements of a cylindrical surface are common with the axis of a datum feature.

Controlled Radius: The limits of the radius tolerance zone must be tangent to the adjacent surfaces, and there can be no reversal in the contour.

Conventional Dimensioning: As used in this text, conventional dimensioning implies dimensioning without the use of geometric tolerancing.

Conventional Tolerancing: As used in this text, conventional tolerancing refers to tolerances related to dimensioning practices without regard to geometric tolerancing.

Coplanar Profile Tolerance: A tolerance that is used when it is desirable to treat two or more separate surfaces, that lie on the same plane, as one surface.

Coplanar Surfaces: Two or more surfaces on a part that are on the same plane.

Cylindricity: A form tolerance identified by a radius tolerance zone establishing two perfectly concentric cylinders that the actual surface must lie within.

Cylindricity Geometric Tolerance: A tolerance that establishes the control of cylindricity.

D

Datum: Datums are considered theoretically perfect surfaces, planes, points, or axes. A datum is assumed to be exact.

Datum Axis: The center axis when a cylindrical datum is featured.

Datum Center Plane: The plane that splits a symmetrical feature such as a slot or tab.

Datum Dimensioning: Each dimension stands alone and is not dependent on the previous dimension. These dimensions are all located from datums.

Datum Feature: The *actual* feature of the part that is used to establish the datum.

Datum Feature Simulators: Simulators that are used to contact the datum features and establish what are known as the simulated datums. These are used for layout purposes.

Datum Feature Symbol: A symbol that is placed on the drawing to identify the features of the object that are specified as datums and referred to as datum features.

Datum Plane: The theoretically exact plane established by the true geometric counterpart of the datum feature.

Datum Precedence: The precedence that is established by the order of placement in the feature control frame. The datums appear in the following order of precedence: primary, secondary, and tertiary.

Datum Reference Frame: The "frame" created by three datum features that are perpendicular to each other and used for layout purposes.

Datum Target: These targets specify areas of contact on a part, points, or lines, that establish datums when it is not possible to use a surface.

Datum Target Line: A target line that is indicated by the target point symbol "X" on the edge view of the surface and by a phantom line on the surface view.

Datum Target Symbol: A symbol that is drawn as a circle with a horizontal line through the center. This symbol is used to identify the characteristics of a datum target.

Diameter: The distance across a circle measured through the center.

Digitizer: An electronic input device that allows data to be entered into the computer by pointing using a pointing device.

Digitizer Cursor: A digitizer cursor is an input device that is hand-held. The screen cursor is a small box or lines crossing on the monitor screen that indicates the current position. The digitizer cursor is the most common pointing device.

Digitizer Tablet Menu: A printed menu displaying commands that are used for specific applications such as mechanical or architectural drafting.

Dimension: A numerical value indicated on a drawing and in documents to define the size, shape, location, geometric characteristics, or surface texture of a feature.

E

Equal Bilateral Tolerance: The variation from the specified dimension is the same in both the + and – directions.

Extreme Attitude Variation: Extreme attitude variation is the axis of the hole at an extreme angle inside the positional tolerance zone.

Extreme Form Variation: The variation of the form of the feature between the upper limit and lower limit of a size dimension.

Extreme Positional Variation: Extreme positional variation is the axis of the hole at the extreme side of the position tolerance zone.

F

Feature: The general term applied to a physical portion of a part or object.

Feature Control Frame: Used to define the geometric tolerancing characteristics of a feature. The feature control frame is divided into compartments containing the geometric characteristic symbol in the first compartment followed by the geometric tolerance.

Feature of Size: One cylindrical or spherical surface, or a set of two parallel plane surfaces, each feature being associated with a size dimension.

Feature-relating Control: The lower entry of the feature control frame that specifies the smaller positional tolerance for the individual features within the pattern when used in a composite positional tolerance.

Fixed Fastener: A fastening situation where one of the parts to be assembled contains a threaded hole for a bolt, screw, or stud located in the second part to be assembled.

Flatness Geometric Tolerance: A tolerance where the feature control frame is connected by a leader or an extension line in the view where the surface appears as a line.

Flatness Tolerance Zone: The establishment of the distance between two parallel planes that the surface must lie within.

Floating Fastener: A fastening situation where two or more parts are assembled with fasteners such as bolts and nuts, and all parts have clearance holes for the bolts.

Force Fits (FN): A special type of interference fit characterized by maintenance of constant pressure. This is where two mating parts must be pressed or forced together. Force fits range from FN 1 (light drive) to FN 5 (force fits required in high stress applications).

Form Tolerance: A tolerance that controls the straightness, flatness, circularity, or cylindricity of a geometric shape.

Form and Orientation Tolerance Zone: When used in a composite profile tolerance, the bottom half of the feature control frame establishes the limits of size, form, and orientation of the profile related to the locating tolerance zone.

Free State Condition: The condition of the part after removal of forces applied during manufacturing.

Free State Variation: A variation that is the distortion of a part after removal of forces applied during manufacture.

G

Geometric Characteristic Symbols: Symbols used in geometric dimensioning and tolerancing to provide specific controls related to the form of an object, the orientation of features, the outlines of features, the relationship of features to an axis, or the location of features.

Graphics Tablet: An electronic input device that allows data to be entered into the computer by pointing using a pointing device.

H

High Points: Points of contact between datums and their datum plane simulators.

I

Interference Fits: Fits that require the mating parts to be pressed or forced together under all tolerance conditions.

L

Least Material Condition (LMC): The condition where a feature of size contains the least amount of material within the stated limits.

Light Pen: A light pen is a type of pointing device that allows a drafter to digitize information into the computer by pressing the "pen" on the screen at the desired location.

Limits: Limits of a dimension are the largest and smallest numerical value possible for the feature within the tolerance specified.

Limits Dimensioning: A system of dimensioning where the upper and lower limits of the tolerance are provided, and there is no specified dimension given.

Limits of Size: The amount of variation in size and geometric form of a feature control. This is referred to as "Rule 1" in ASME Y14.5M. The limits of size is the boundary between maximum material condition (MMC) and least material condition (LMC).

Locating Tolerance Zone: The top half of the feature control frame that locates the feature from datums used in a composite profile tolerance.

Location Tolerance: Tolerance used for the purpose of locating a feature from datums or for establishing coaxiality or symmetry. Location tolerances include position, concentricity, and symmetry.

Locational Fits: These fits *only* determine the *location* of mating parts.

M

Material Condition Symbols: Symbols that establish the relationship between the size of the feature within its given dimensional tolerance and the geometric tolerance. Often referred to as modifying symbols.

Maximum Material Condition (MMC): The condition where a feature contains the maximum amount of material within the stated limits of size.

Menu Tablet: An electronic input device that allows data to be entered into the computer by pointing using a pointing device.

Mouse: A device that senses its position on a flat surface by movement of a ball or by reflected light across a grid. When the mouse is moved across a flat surface, the screen cursor also moves. Buttons on the mouse activate specific functions or allow the user to choose from menu items displayed on the screen.

Multiple Datum Reference: A datum reference that is established by two datum features such as an axis established by two datum diameters. When a multiple datum reference is used, all applicable datum reference letters, separated by a dash, are placed in a single compartment after the geometric tolerance.

N

Nominal size: A dimension used for general identification such as stock size or thread diameter.

Nonrigid Parts: Parts that may have dimensional change due to thin wall characteristics.

O

Order of Precedence: The order that the datums appear in the last three compartments of the feature control frame. The primary datum is given first, followed by the secondary datum and then the tertiary datum.

Orientation Geometric Tolerance: A tolerance that controls parallelism, perpendicularity, angularity, or runout as a combination of controls. When controlling orientation tolerances, the feature is related to one or more datum features.

P

Parallelism Geometric Tolerance: A tolerance defined by two parallel planes or cylindrical zones that are parallel to a datum plane or axis where the surface or axis of the specified feature must lie.

Pattern-locating Control: The upper part of the feature control frame that specifies the larger positional tolerance for the pattern of features as a group used in a composite positional tolerance.

Perfect Flatness: The condition of a surface where all of the elements are in one plane.

Perfect Form Boundary: The boundary of size tolerance limits is the true geometric form of the feature. In other words, the limit where perfect form of the feature is required is the perfect form boundary.

Perfect Symmetry: Occurs when the center planes of two or more related symmetrical features line up.

Perpendicularity Tolerance: A tolerance that is established by a specified geometric tolerance zone made up of two parallel planes or cylindrical zones that are a basic 90° to a given datum plane or axis where the actual surface or axis must lie.

Plus-minus Dimensioning: A system of dimensioning that provides a nominal dimension and an amount of allowable variance from that dimension. May appear similar to 24.5±0.5.

Pointing Devices: A variety of instruments that are attached to the digitizer or computer terminal. These devices allow the user to control an on-screen cursor.

Polar Coordinate Dimensioning: Dimensioning where angular dimensions are combined with other dimensions to locate features from planes, centerlines, or center planes.

Positional Tolerance: A tolerance that is used to define a zone where the center, axis, or center plane of a feature of size is permitted to vary from true position.

Primary Datum Plane: A plane that must be established by at least three points on the primary datum surface. This datum plane is first in the precedence of datums.

Profile: The outline of an object represented either by an external view or by a cross-section through the object.

Profile of a Line Tolerance: A two-dimensional or cross-sectional geometric tolerance that extends along the length of the feature.

Profile of a Surface: A geometric tolerance that controls the entire surface of a feature or object as a single entity.

Profile Tolerance: A uniform boundary along the true profile that the elements of the surface must lie within.

Projected Tolerance Zone: A tolerance zone that is recommended when variations in perpendicularity of threaded or press-fit holes could cause the fastener to interfere with the mating part. The projected tolerance zone is established at true position and extends away from the primary datum at the threaded feature.

R

Radius: The distance from the center of a circle to the outside.

Radius Dimension: The limits of the radius tolerance zone are not required to be tangent to the adjacent surfaces and reversals in the contour are permitted. See *"Controlled Radius"* for a way of defining a tighter specification.

Rectangular Coordinate Dimensioning: Dimensioning where linear dimensions are used to locate features from planes, centerlines, or center planes.

Reference Dimension: A dimension, usually without a tolerance, used for information purposes only. This dimension is often a repeat of a given dimension or established from other values shown on the drawing. A reference dimension is shown on a drawing enclosed in parentheses.

Regardless of Feature Size (RFS): The term used to indicate that a geometric tolerance or datum reference applies at any increment of size of the feature within its size tolerance. RFS is assumed for all geometric tolerances unless otherwise specified.

Running and Sliding Fits (RC): These fits provide a running performance with suitable lubrication allowance. Running fits range from RC1 (close fits) to RC9 (loose fits).

"Rule 1," ASME Y14.5M: The amount of variation in size and geometric form of a feature control. The boundary between maximum material condition (MMC) and least material condition (LMC).

"Rule 2–All Applicable Geometric Tolerances," ASME Y14.5M: States that RFS applies, with respect to the individual geometric tolerance and/or datum reference, when no material condition symbol is specified.

Runout: A combination of controls including: the control of circular elements of a surface; the control of the cumulative variations of circularity, straightness, coaxiality, angularity, taper, and profile of a surface; and control of variations in perpendicularity and flatness.

S

Secondary Datum Plane: A plane that must be located by at least two points on the related secondary datum surface. The secondary datum plane is the second datum in the order of precedence.

Shrink Fits: A special type of interference fit characterized by maintenance of constant pressure. (Also known as force fits.) This is where two mating parts must be pressed or forced together.

Simulated Datum: The surface or axis established by the inspection equipment such as a surface plate or inspection table. A simulated datum is used for measuring or inspection purposes.

Simulated Datum Axis: The axis of a perfect cylindrical inspection device that contacts the datum feature surface.

Simulated Datum Center Plane: The center plane of a perfect rectangular inspection device that contacts the datum feature surface.

Single Composite Pattern: A group of features that are located relative to common datum features not subject to size tolerance, or to common datum features of size specified on an RFS basis.

Specific Area Flatness: A tolerance used where a large cast surface must be flat, but it is possible to finish only a relatively small area, rather than an expensive operation of machining the entire surface. Defines a portion of a surface where the tolerance applies.

Specified Dimension: The part of the dimension from where the limits are calculated.

Spherical Radius: The distance from the surface of a spherical feature to its center point.

Stacking: The occurrence of the tolerance of each individual dimension builds on the next. Also known as tolerance buildup.

Statistical Process Control (SPC): A method of monitoring a manufacturing process by using statistical signals to either leave the process alone, or change it as needed to maintain the quality intended in the dimensional tolerancing.

Statistical Tolerance: To assign tolerances to related dimensions in an assembly based on the requirements of statistical process control (SPC).

Stock Size: A commercial or premanufactured size, such as a particular size of square, round, or hex steel bar.

Straightness: The measure of how closely an element of a surface or an axis is to a perfect straight line.

Straightness Per Unit: A measure of a part or feature in conjunction with a straightness specification over the total length.

Straightness Tolerance: A specified zone where the required surface element or axis must lie.

Surface Geometric Control: The connection of the feature control frame with either a leader to the surface of the object or feature, or extended from an extension line from the surface of the object or feature.

Surface Straightness Tolerance: The connection of the feature control frame to the surface with a leader, or the connection of the feature control frame to an extension line in the view where the surface to be controlled is shown as an edge.

Symbols: Symbols represent specific information that would otherwise be difficult and time consuming to duplicate in note form.

Symmetry: Symmetry is a center plane relationship of the features of an object.

Symmetry Geometric Tolerance: A zone where the median points of opposite symmetrical surfaces align with the datum center plane.

T

Tangent Plane: A plane that is theoretically exact and is established by the true geometric counterpart of the feature surface.

Tangent Plane Symbol: A tangent plane symbol is used when a plane contacting the high points of surface must be within the specified geometric tolerance zone.

Template: Typically made of thin plastic and containing different symbols for use in geometric dimensioning and tolerancing that the drafter can trace.

Tertiary Datum Plane: A plane that must be located by at least one point on the related tertiary datum surface. The tertiary datum plane is third in the order of precedence.

Tolerance: The total amount that a specific dimension is permitted to vary.

Tolerance Buildup: The occurrence of the tolerance of each individual dimension builds on the next. Also known as stacking.

Total Runout: A tolerance that blankets the surface to be controlled and provides a combined control of a surface element. Total runout is used to control the combined variations of circularity, straightness, coaxiality, angularity, taper, and profile when applied to surfaces around a datum axis.

Trackball: Can be described as an "up-side-down" mouse, except it remains stationary. The user rotates a ball that is located on the top of the unit. That movement is in turn translated by the computer into cursor movement.

Transition Fits: These fits may result in either a clearance fit or an interference fit due to the range of limits between mating parts.

True Geometric Counterpart: A simulated datum that is the processing equipment used for layout and measurement.

True Position: A theoretically exact location for a feature.

True Profile: The actual desired shape of the object that is the basis of the profile tolerance.

Two Single Segment Feature Control Frame: Similar to composite positional tolerance, except the feature control frame has two position symbols, each displayed in a separate compartment, and a two datum reference in the lower half of the feature control frame controlling the orientation and alignment with the pattern locating control. This provides a tighter relationship of the pattern than a composite positional tolerance.

U

Unequal Bilateral Tolerance: The variation from the specified dimension is not the same in both directions.

Unilateral Profile: The entire profile tolerance zone is on one side of the true profile.

Unilateral Tolerance: This tolerance permits the increase or decrease in only one direction from the specified dimension.

Unit Flatness: Unit flatness may be specified when it is desirable to control the flatness of a given surface area as a means of controlling an abrupt surface variation within a small area of the feature.

V

Variable Dimensions: Dimensions that are labeled with letters that correlate with a chart where different options are shown.

Virtual Condition: The combined maximum material condition and geometric tolerance (Virtual Condition = MMC + Geometric Tolerance).

Z

Zero Positional Tolerance: This tolerance at MMC is used with the design of the maximum material condition of clearance holes at virtual condition.

Index